Mathematics for Elementary Teachers
Preliminary Edition

Volume 1
Numbers and Operations

Sybilla Beckmann

University of Georgia

Addison
Wesley

Boston San Francisco New York
London Toronto Sydney Tokyo Singapore Madrid
Mexico City Munich Paris Cape Town Hong Kong Montreal

Reproduced by Addison-Wesley from electronic files supplied by the author.

Copyright © 2003 Pearson Education, Inc.

ISBN   0-321-12980-6

1 2 3 4 5 6 7 8 9 10 VHG 04 03 02 01

This book is dedicated to Will, Joey, and Arianna

Contents

Features

Text: The text introduces and discusses the concepts and principles.

Exercises and Answers to Exercises: Almost every section has exercises that are followed by detailed answers to these exercises. Students should try to solve the exercises on their own first, consulting the text for the concepts and principles as needed. After solving, or attempting to solve, the exercises, students should consult the answers to the exercises. The answers serve as examples of good explanations.

Problems: Almost every section has problems for which answers are not provided.

Companion Book of Class Activities: There is a companion book of class activities which are designed to be used in class. In the text, you will find references to the class activities. The reference to an activity is placed at a point in the text where it makes sense for students to work on the activity.

Preface

I wrote this book to help elementary teachers develop a deep understanding of the mathematics that they will teach. It is easy to think that the mathematics of elementary school is simple, and that it shouldn't require college-level study in order to teach it well. But to teach mathematics well, teachers must know more than just *how* to carry out basic mathematical procedures, they must be able to explain *why* mathematics works the way it does. Knowing *why* requires a much deeper understanding than knowing *how*. For example, it is easy to multiply fractions—multiply the tops and multiply the bottoms—but *why* do we multiply fractions that way? After all, when we *add* fractions we can't just add the tops and add the bottoms. The reasons why are not obvious: they require study. By learning to explain why mathematics works the way it does, teachers will learn to make sense of mathematics. I hope they will carry this sense-making into their own future classrooms.

This book focuses on "explaining why". Prospective elementary teachers will learn to explain why the standard procedures and formulas of elementary mathematics are valid, why non-standard methods can also be valid, and why other seemingly plausible ways of reasoning are not correct. The book emphasizes key concepts and principles, and it guides prospective teachers to give explanations that draw on these key concepts and principles. I hope that this will help teachers organize their knowledge around the key concepts and principles of mathematics, so that they will be able to help their students do likewise.

A number of problems and activities examine common misconceptions. I hope that by having studied and analyzed these misconceptions, teachers will be able to explain to their students why an erroneous method is wrong instead of just saying "you can't do it that way".

In addition to knowing how to explain mathematics, prospective teachers should also know how mathematics is used. Therefore I have included various examples of how we use mathematics, such as using proportions to compare prices across years with the Consumer Price Index, using visualization skills to explain the phases of the Moon, using angles to explain why spoons reflect upside down, and using spheres to explain how the Global Positioning System works.

I believe that this book is an excellent fit for the recommendations of the Conference Board of the Mathematical Sciences (CBMS) on the mathematical preparation of teachers, and that it will help prepare teachers to teach to the Principles and Standards of the National Council of Teachers of Mathematics (NCTM).

In writing this book I have benefited from much help and advice. I would like to thank Malcolm Adams, Ed Azoff, David Benson, Cal Burgoyne, Denise Mewborn, Ted Shifrin, Shubhangi Stalder, Gale Watson, and Paul Wenston for many helpful comments on an earlier drafts. I also owe special thanks to Kevin Clancey, John Hollingsworth, and Tom Cooney for their advice and encouragement. I thank my family Will, Joey and Arianna Kazez for their patience, support, and encouragement. My children long ago gave up asking, "Mom, are you still working on that book?", and are now convinced that books take forever to write.

Sybilla Beckmann

Athens, Georgia
May, 2002

Chapter 1

Problem Solving

In this introductory chapter, we will see why problem solving is a central part of mathematics, and we will study a simple but sensible guideline for solving problems. When you solve a problem, you should be able to explain why your solution is valid. Thus the second part of this chapter discusses features of a good explanation. A main theme of this book is *explaining why*: why are the familiar procedures and formulas of elementary mathematics valid? why is a student's mistake incorrect? why is a different way of carrying out a calculation often perfectly correct?

1.1   Solving Problems

The main reason for learning mathematics is to be able to solve problems. Mathematics is a powerful tool that can be used to solve a vast variety of problems in technology, in science, in business and finance, in medicine, in daily life. The list could stretch on and on, and is limited only by human ingenuity. But solving problems is not only the most important *end* of mathematics, it is also a *means* for learning mathematics. Mathematicians have long known that good problems can deepen our thinking about mathematics, good problems can guide us to new ways of using mathematical techniques we have learned, good problems can help us recognize connections between topics in mathematics, and good problems force us to confront misconceptions we may hold in mathematics. By working on good problems, we learn mathematics better. This is why the National Council of Teachers of Mathematics advocates that instructional programs for grades pre-K through 12 focus on problem solving.

The National Council of Teachers of Mathematics (NCTM)—an organization that is over 75 years old and has over 100,000 members—advocates a mathematics education of the highest quality for all students (see `nctm.org`). To this end, the NCTM has produced the document *Principles and Standards for School Mathematics* [40], which includes the following standard on problem solving:

> Instructional programs from prekindergarten through grade 12 should enable all students to—

- solve problems that arise in mathematics and in other contexts;

- apply and adapt a variety of appropriate strategies to solve problems;

- monitor and reflect on the process of mathematical problem solving.

Polya's Four Problem-Solving Steps

In 1945, the mathematician George Polya presented a four step guideline for solving problems in his book *How to Solve It* [49]. These four steps are simple and sensible, and have helped many students improve their problem-solving abilities:

1. Understand the problem.

2. Devise a plan.

3. Carry out the plan.

4. Look back.

The first step, *understanding the problem*, is the most important. Although it is obvious that if you don't understand a problem, you won't be able to solve it, it is easy to rush headlong into a problem and to try to do "something like what we did in class" before you think about what the problem is asking. So *slow down* and read problems carefully. In some cases, drawing a diagram or a picture can help you understand the problem.

Try to be creative and flexible in the second step, when *devising a plan*. There are many different types of plans for solving problems. In devising a plan, think about what information you know, what information you are looking for, and how to relate these pieces of information. The following are a few common types of plans:

- Draw a diagram or a picture.

- Guess and test—make a guess, try it out. Be prepared to try again. Try to use the results from your first guess to guide you.

- Look for a pattern.

- Solve a simpler problem or problems first—maybe this will help you see a pattern you can use.

- Use direct reasoning.

- Use a variable, such as x.

The third step, *carrying out the plan*, is often the hardest part. If you get stuck, try to modify your plan, or try a new plan. It often helps to cycle back through steps 1 and 2. Monitor your own progress: if you are stuck, is it because you haven't tried hard enough to make your plan work, or is it time to try a new plan? But don't give up too easily. Students sometimes think they can only solve a problem if they've seen one just like it before, but this is not true. Your common sense and your natural thinking abilities are powerful tools that will serve you well if you make an effort to use them.

The fourth step, *looking back*, gives you an opportunity to catch mistakes. Check to see if your answer is plausible. For example, if the problem was to find the height of a telephone pole, then answers such as 2.3 feet or 513 yards are clearly ridiculous—there must be a mistake somewhere. *Looking back* also gives you an opportunity to make connections: have you seen this type of answer before? what did you learn from this problem? could you use these ideas in some other way? is there another way to solve the problem? When you *look back*, you have an opportunity to learn from your own work.

Why Do I Need to Solve Problems Myself When the Teacher Could Just Tell Me How?

Students sometimes wonder why they need to try to solve problems themselves. Why can't the teacher just tell the students how to solve the problem? Of course, teachers should show their students how to solve many problems. But sometimes teachers should step back and *guide* their students, helping their students to use fundamental concepts and principles to *figure out* how to solve a problem. Why? Because *the process of grappling with a problem can help students make sense of the underlying concepts and principles*. Therefore, teachers who are too quick to tell students how to solve problems may actually rob their students of valuable learning experiences.

When you solve problems yourself, you will inevitably get stuck some of the time, and you will try things that don't work. This may seem unproductive, but it is not. Even if you are not able to solve a problem, if you get stuck and try hard to get unstuck you have primed yourself to *better understand a solution presented by someone else.* Has it ever happened to you that a teacher gave a detailed explanation of something and you had no idea what the teacher was talking about? If you have genuinely tried to solve a problem, you will be much more likely to understand a solution later on, even if you couldn't solve the problem right away.

Stop for a moment and think about something that you know well. This could be an academic topic, or it could be from sports, music, religion, or any aspect of your life. Chances are that this "something that you know well" is something that you have thought deeply about, something that you have grappled with, maybe even struggled with. You probably learned a lot about it from others, but it is likely that you have put effort into *making sense of it for yourself.* The process of making sense of things for yourself is the essence of education—use it in mathematics, and do not underestimate its power.

Class Activity 1A: A *Clinking Glasses* Problem

Class Activity 1B: Problems About Triangular Numbers

Class Activity 1C: What is a Fair Way to Split the Cost?

1.2 Explaining Solutions

After solving a problem, the natural next step is to explain why your solution is valid. Why does your method work? Why does it give the correct answer to the problem? *Explaining why* is the main emphasis of this book. As a teacher, you will need to explain why the mathematics you are teaching works the way it does. But there is an even more compelling reason for emphasizing explanations: *when you try to explain something to someone else, you clarify your own thinking, and you learn more yourself.* When you try to explain a solution, you may find that you don't understand it as well as you thought. This is valuable: it is an opportunity to learn more, to uncover an error, or clear up a misconception. Even if you understood the solution well, you will understand it better after explaining it. This is why the National Council of Teachers of Mathematics (NCTM) includes communication as one of their standards in [40]:

Instructional programs from prekindergarten through grade 12 should enable all students to—

- organize and consolidate their mathematical thinking through communication;

- communicate their mathematical thinking coherently and clearly to peers, teachers, and others;

- analyze and evaluate the mathematical thinking and strategies of others;

- use the language of mathematics to express mathematical ideas precisely.

Communicating about mathematics gives both children and adults an opportunity to make sense of mathematics. According to NCTM, [40, p. 56]:

> From children's earliest experiences with mathematics, it is important to help them understand that assertions should always have reasons. Questions such as "Why do you think it is true?" and "Does anyone think the answer is different, and why do you think so?" help students see that statements need to be supported or refuted by evidence.

When we communicate about mathematics in order to explain and convince, we must use *reasoning*. Logical reasoning is the very essence of mathematics. In mathematics, everything but the fundamental starting assumptions has a reason, and the whole structure of mathematics is built up by reasoning. Therefore NCTM includes a standard on reasoning and proof [40]:

Instructional programs from prekindergarten through grade 12 should enable all students to—

- recognize reasoning and proof as fundamental aspects of mathematics;

- make and investigate mathematical conjectures;

- develop and evaluate mathematical arguments and proofs;

- select and use various types of reasoning and methods of proof.

What is an Explanation in Mathematics?

What qualifies as an explanation? The answer depends on the context. In mathematics, we seek particular kinds of explanations: ones that use logical reasoning and are based on starting assumptions that are either explicitly stated or are assumed to be understood by the reader.

Explanations can be different in different areas of knowledge. There are many different kinds of explanations—even of the same phenomenon. For example, consider the question: why are there seasons?

The simplest answer is: because that's just the way it is. Every year that we live, we observe the passing of the seasons, and we expect to see the cycle of spring, summer, fall, and winter continue indefinitely (or at least for a very long time). The cycle of seasons is an observed fact that has been recorded throughout history. We could stop there, but when we ask why there are seasons, we are searching for a deeper explanation.

Figure 1.1: Spring

A poetic explanation for the seasons might refer to the many cycles of birth, death, and rebirth around us. In our experiences, nothing remains

Figure 1.2: Summer

Figure 1.3: Fall

Figure 1.4: Winter

unchanged forever, and most everything is part of a cycle. The cycle of the seasons is one of the many cycles that we observe.

The ancient Greeks explained the seasons with the story of Persephone and her mother, Demeter, who tends the earth. Pluto, god of the underworld stole Persephone to become his bride. But Persephone and Demeter were heartbroken. In the end, a compromise was arranged: Persephone must spend 6 months of each year in the underworld with Pluto, and the remaining 6 months of the year with Demeter, above ground. When Persephone is in the underworld, Demeter is sad and does not tend the earth. Leaves fall from the trees, flowers die, and it is fall and winter. When Persephone returns, Demeter is happy again and tends the earth. Leaves grow on the trees, flowers bloom, and it is spring and summer. This is a beautiful story, but we can still wish for more.

Modern scientists explain the reason for the seasons by the tilt of the earth's axis relative to the plane in which the earth travels around the sun. When the northern hemisphere is tilted toward the sun, it is summer there; when it is tilted away from the sun, it is winter. Perhaps this settles the matter, but a seeker could still ask for more. Why are the earth and sun positioned the way they are? Why does the earth revolve around the sun and not fly off alone into space? These questions can lead again to poetry, or to the spiritual, or to further physical theories. Maybe they lead to an endless cycle of questions.

Writing Good Explanations

Oral explanations and discussions can be helpful, but writing an explanation pushes you to polish, refine, and clarify your ideas. This is as true in mathematical writing as it is in any other kind of writing, and it is true at all levels. Some elementary school teachers have successfully integrated mathematics and writing in their classrooms, and used writing to help their students develop their understanding of mathematics.

Like any kind of writing, it takes work and practice to write good mathematical explanations. When you solve a problem, do not attempt to write the final draft of your solution right from the start. Use scrap paper to work on the problem and collect your ideas. Then write up your solution as part of the *looking back* portion of problem solving. Think of your explanation as an essay. As with any essay that aims to convince, what counts is not only factual correctness but also persuasiveness, explanatory power, and clarity of expression. In mathematics, we persuade by giving a thorough, logical argument, in which chains of logical deductions are strung together connecting the starting assumptions to the desired conclusion.

Good mathematical explanations are thorough: they should not have gaps that involve leaps of faith. On the other hand, a good explanation should not belabor points that are well-known to the audience and are not central to the explanation. For example, a college-level explanation that involves the calculation of $356 \div 7$ need not explain how the calculation is carried out, unless a description of the calculation is important for the explanation. Unless your instructor tells you otherwise, assume that you are writing your explanations for your classmates.

On the next page you will find characteristics of good mathematical explanations. When you write an explanation, check if it has these characteristics. The answers to the exercises at the ends of the sections give you many examples of good mathematical explanations. The more you work at writing explanations, and the more you *think about*, and analyse what makes good explanations, the better you will be able to write explanations, and *the better you will understand the mathematics involved*.

Characteristics of Good Explanations in Mathematics

1. The explanation is factually correct, or nearly so, with only minor flaws (for example, a minor mistake in a calculation).

2. The explanation addresses the specific question or problem that was posed. It is focused, detailed, and precise. There are no irrelevant or distracting points.

3. The explanation is clear, convincing, and logical. A clear and convincing explanation is characterized by the following:

    (a) The explanation could be used to teach another (college) student, possibly even one who is not in the class.

    (b) The explanation could be used to convince a skeptic.

    (c) The explanation does not require the reader to make a leap of faith.

    (d) Key points are emphasized.

    (e) If applicable, supporting pictures, diagrams, and/or equations are used appropriately and as needed.

    (f) The explanation is coherent.

    (g) Clear, complete sentences are used.

Sample Problems and Solutions

The solutions to the problems below provide you with some sample explanations. In addition to these explanations, many of the answers to the exercises at the end of the sections throughout this book provide additional examples of explanations.

A Puzzle Problem and Three Solutions

Puzzle Problem: There are 9 coins all together, some of them pennies, some nickels and some dimes. If you collect up all the pennies and nickels, there are 7. If you collect up all the nickels and dimes, there are 5. How many of the 9 coins are pennies, how many are nickels, how many are dimes?

   This problem can be solved in many different ways.

   Solution 1: If there are 9 pennies, nickels and dimes, and 7 are pennies and nickels, then 2 must be dimes. Since the nickels and dimes together make 5 coins, and since there are 2 dimes, therefore there must be 3 nickels. But once again: there are 7 pennies and nickels, so since 3 of them are nickels, 4 must be pennies. So there are 2 dimes, 3 nickels, and 4 pennies. As a check, add up the total number of coins: $2 + 3 + 4 = 9$, which agrees with our starting data. Also, there are $3 + 4 = 7$ pennies and nickels, and there are $3 + 2 = 5$ nickels and dimes, so these check too.

   Solution 2: There are 7 pennies and nickels and there are 5 nickels and dimes, so if we add 7 and 5, that will count all the nickels twice, and all the pennies and dimes once. Therefore $7 + 5 = 12$ represents the total number of coins plus the the number of nickels. Since the total number of coins is 9, therefore the number of nickels must be $12 - 9 = 3$. But we know that there are 7 pennies and nickels together, so since there are 3 nickels, then there must be 4 pennies. We also know that there are 5 nickels and dimes together, so since there are 3 nickels, there must be 2 dimes. So there are 3 nickels, 4 pennies and 2 dimes. As a check, add up the total number of coins: $3 + 4 + 2 = 9$ which agrees with our starting data. Also, there are $3 + 4 = 7$ pennies and nickels, and there are $3 + 2 = 5$ nickels and dimes, so these check too.

   Solution 3: Let P be the number of pennies, N the number of nickels, and D the number of dimes. The problem tells us that:

$$P + N + D = 9,$$

$$P + N = 7,$$

$$N + D = 5.$$

From the second and third equations,

$$P = 7 - N,$$

$$D = 5 - N.$$

Substituting for P and D in the first equation:

$$(7 - N) + N + (5 - N) = 9,$$

therefore

$$-N + 12 = 9,$$

so that

$$12 - 9 = N,$$

so $N = 3$. Since $P + N = 7$, therefore $P + 3 = 7$, so $P = 4$. Since $N + D = 5$ and $N = 3$, therefore $3 + D = 5$ and $D = 2$. So there are 3 nickels, 4 pennies and 2 dimes. Since $3 + 4 + 2 = 9$ this checks with the initial information that there are 9 coins. Also, there are $3 + 4 = 7$ pennies and nickels, and there are $3 + 2 = 5$ nickels and dimes, so these check too.

Class Activity 1D: Who Says You Can't Do Rocket Science?

Exercises for Section 1.2 on Explaining Solutions

1. Joanne and Sahar make plans to go to ten concerts together. Sahar buys 7 pairs of tickets which cost a total of $350. Joanne buys 3 pairs of tickets which cost a total of $63. Who owes whom how much?

2. The shape in Figure 1.5 is called an "arithmogon". The problem is to find numbers to put in the circles. These numbers should be chosen so that when you add numbers in circles connected by a line segment, you get the number in the box next to the line segment.

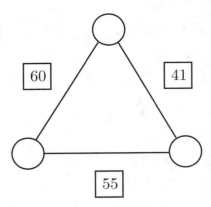

Figure 1.5: An Arithmogon

3. If there are 25 people at a party and everyone "clinks" glasses with everyone else, how many "clinks" will there be? Describe *two different ways* to solve this problem.

4. Explain how you can use what you learned from Class Activity 1A on *clinking glasses* (and/or the previous exercise) to help you determine the 500th triangular number.

Answers to Exercises for Section 1.2 on Explaining Solutions

Most of the answers here, as well as in other sections, include an explanation. Please realize that each explanation is generally only one of many possible valid explanations.

1. Joanne owes Sahar $143.50. To figure this out, notice that in combination, Joanne and Sahar have spent $350 + $63 = $413. Therefore, in the end, each woman should spend half that amount, namely $206.50. Because Joanne already spent $63, she needs to spend an additional $206.50 − $63 = $143.50.

2. See Figure 1.6. One way to solve this problem is to make an initial guess for the three numbers, picking them so that two of the pairs add up correctly. Then revise your guess, maintaining the two correct sums. You can also solve this using algebra.

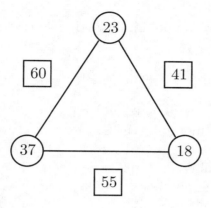

Figure 1.6: An Arithmogon

3. One way to solve this problem is to add

$$24 + 23 + 22 + \ldots + 4 + 3 + 2 + 1 = 300$$

to find the total number of "clinks." To see why, imagine the 25 people lined up in a row. The 25th person in the row could go down the row and clink glasses with all other 24 people (and then go stand in the corner). The 24th person in the row could go down the row and clink glasses with the other 23 people left in the row (and then go stand in the corner too). The 23rd person could clink with the remaining 22, and so on, until the 2nd person clinks with the first. Counting up all the clinks, you get the sum $24 + 23 + \ldots 2 + 1$ above.

Another way to solve this problem is to calculate

$$(25 \times 24) \div 2 = 300.$$

This method is valid because each of the 25 people "clinks" with 24 other people. This makes for 25×24 clinks—except that each clink has been counted *twice* this way. Why? Because when Anna clinks with Beatrice it's the same as Beatrice clinking with Anna, but the 25×24 figure counts those as two separate clinks. So to get the correct count, divide 25×24 by 2.

4. Imagine adding up the number of dots in the 500th triangle by starting with the bottom row and working your way up, one row at a time.

According to this way of thinking, the 500th triangular number must be equal to

$$500 + 499 + 498 + 497 + \ldots + 3 + 2 + 1.$$

It would be pretty tedious to add up all those numbers! Fortunately, there's a quicker method. To see how that works, think about the explanation for the previous exercise. There are two different ways of counting the number of clinks when there are 25 people at a party. Since it's the same number of clinks, no matter which way you count them, therefore

$$24 + 23 + 22 + \ldots + 4 + 3 + 2 + 1 = (25 \times 24) \div 2.$$

But now think about doing the same analysis, but when there are 501 people at a party. By the same reasoning as above, one way of counting the number of clinks is

$$500 + 499 + 498 + 497 + \ldots + 3 + 2 + 1$$

while another way is

$$(501 \times 500) \div 2.$$

But there must be the same number of clinks either way you count them, therefore

$$500 + 499 + 498 + 497 + \ldots + 3 + 2 + 1 = (501 \times 500) \div 2.$$

Even though it appears to be entirely unlrelated, we can apply this to the triangular number problem and conclude that there are

$$(501 \times 500) \div 2 = 125250$$

dots in the 500th triangle.

Problems for Section 1.2 on Explaining Solutions

1. Denise, Sharon, and Hillary go out to dinner and have a meal that comes to a total of $45 (including tax and tip). They decide to split the bill evenly, so each woman owes $15. Sharon wants to put the total on her credit card and have the other two give her cash or a check.

Hillary only has one check, three $1 bills and a $5 bill. Denise only has a $10 bill and a $20 bill. Explain how the women can settle among themselves so that the waiter only has to take Sharon's credit card for the total amount (and so that nobody owes anything). Explain your reasoning.

2. The Bradley Elementary School cafeteria has twelve different lunches that they can prepare for their students. Five of these lunches are "reduced fat". On any given day, the cafeteria offers a choice of two lunches. How many different pairs of lunches, where one choice is "regular" and the other is "reduced fat" is it possible for the cafeteria to serve? Explain your answer clearly.

3. Ms. Jones, an elementary school teacher, has 24 students. Every day she picks two students to be her "special helpers" for the day. How many pairs of students are there in the class? If the school year is 190 days long, will there be enough days for every pair of children to be the "special helpers"? (Notice that the problem asks about every *pair* of children, not every individual child.)

4. (a) Explain how to determine the sum

$$1 + 2 + 3 + \cdots + 100$$

using ideas from the *clinking glasses* class activity (Class Activity 1A).

(b) Another way to determine the sum

$$1 + 2 + 3 + \cdots + 100$$

is to add the following "vertically":

1	+	2	+	3	+	4	+	$\cdots$	+	99	+	100
+100	+	99	+	98	+	97	+	$\cdots$	+	2	+	1

Explain how to use this approach to determine the sum $1 + 2 + 3 + \cdots + 100$.

5. An auditorium has 50 rows of seats. The first row has 20 seats, the second row has 21 seats, the third row has 22 seats, and so on. How many seats are there all together? Explain your reasoning.

6. A candy factory has a large vat into which they pour chocolate and cream. Each ingredient is poured into the vat from its own special hose, and each ingredient comes out of its hose at a constant rate. Workers at the factory know that it takes 20 minutes to fill the vat with chocolate from the chocolate hose, and it takes 15 minutes to fill the vat with cream from the cream hose. If workers pour both chocolate and cream into the vat at the same time (each coming full tilt out of its own hose), how long will it take to fill the vat? Before you find an exact answer to this problem, find an approximate answer, or find a range, such as "between ... and ... minutes." Explain your reasoning clearly.

7. The Joneses and the Smiths go on a trip together. There are four people in the Jones family and six in the Smith family. They take a ferry boat to get to their destination. This costs $12 per person and the Joneses pay for it. The Smiths pay for dinner at a lodge that costs $15 per person. If the Joneses and the Smiths want to divide up the costs fairly, then who owes whom how much? Explain!

8. Jane, Kayla and Mandelite want to give their friend Nadine a potted plant for her birthday. Jane spends $34 on a pot, Kayla spends $3 on potting soil and Mandelite buys a plant for $24. If Jane, Kayla and Mandelite want to share their costs equally then who should give whom how much? Explain!

9. The country of Taxalot has a different method of taxing income. If you make $1000 per year your tax rate is 1%. For each additional $1000 you make per year, your tax rate goes up by 1%, so that, for example, if you made $24,000 per year you would pay 24% in taxes. If you could set your salary anywhere between $0 and $100,000, what would you want it to be and why?

10. Suppose that a fireworks rocket is shot up into the air so that after 4 seconds, it reaches a height of 1600 feet. Two seconds after being shot into the air, is the fireworks rocket 800 feet high, less than 800 feet high, or more than 800 feet high? Explain your reasoning.

Chapter 2

Numbers

This chapter is about how we commonly write and represent numbers, how we interpret the meaning of written numbers, and how we compare sizes of numbers. Writing, representing, interpreting, and comparing numbers are fundamental skills that are essential not only for the study of mathematics itself, but for effective functioning in today's society. For example, when we go to a store, our knowledge of how numbers are represented, interpreted, and compared affects our ability to make informed choices: do I have enough money to buy this item? is this brand a better buy than that brand? did the clerk really give me a 35% discount? When we open a newspaper, we may read about a federal debt in trillions of dollars or a population increase in our county of 8%. The more we know about numbers and the more adept we are at dealing with them, the better we can think critically about the information that comes our way.

A good grasp of how to work effectively with numbers, and mathematics in general, is also an essential part of preparation for the workplace. As a teacher, you will be responsible for preparing your students not only to be empowered citizens, but also to be ready for productive employment. One hundred years ago, knowing how to add and subtract numbers rapidly was a marketable skill; today that is not enough—any calculator can do as much and more. One hundred years ago a large proportion of jobs were unskilled; today, unskilled work is a shrinking portion of our economy. Much of today's work requires an ability to think critically and to do more than just routine tasks. Some of the fastest growing segments of our economy, such as health care, high technology, and finance, require an increasingly sophisticated understanding of how to work with numbers in particular, and mathematics in general. Schooling that only emphasizes the routine, mechanical aspects of working with numbers will leave students unprepared for the kind of critical thinking that will provide them with a wide range of opportunities for good employment. But schooling that asks students to interpret, analyze, and explain will equip students not only with skills for a wide range of choices for employment, but even more importantly, with the most empowering asset of all: that of deep knowledge and understanding. So that you can empower and guide your students, you must first empower yourself with a deeper understanding of numbers, arithmetic, and mathematics in general.

The National Council of Teachers of Mathematics (NCTM) has developed a collection of standards for the mathematics that children from grades pre-Kindergarten through 12 should know and be able to do, [40]. Concerning numbers, the NCTM recommends that all children in grades Pre-K to 2 should:

- count with understanding and recognize "how many" in sets of objects;

- use multiple models to develop initial understandings of place value and the base-ten number system;

- develop understanding of the relative position and magnitude of whole numbers and of ordinal and cardinal numbers and their connections;

- develop a sense of whole numbers and represent and use them in flexible ways, including relating, composing, and decomposing numbers;

- connect number words and numerals to the quantities they represent, using various physical models and representations;

- understand and represent commonly used fractions, such as 1/4, 1/3, and 1/2.

For grades 3 to 5 the NCTM recommends that all students should:

- understand the place-value structure of the base-ten number system and be able to represent and compare whole numbers and decimals;

- recognize equivalent representations for the same number and generate them by decomposing and composing numbers;

- develop understanding of fractions as parts of unit wholes, as parts of a collection, as locations on number lines, and as divisions of whole numbers;

- use models, benchmarks, and equivalent forms to judge the size of fractions;

- recognize and generate equivalent forms of commonly used fractions, decimals, and percents;

- explore numbers less than 0 by extending the number line and through familiar applications;

- describe classes of numbers according to characteristics such as the nature of their factors.

This chapter is devoted to deepening your own understanding of most of these concepts (except for the topics concerning factors and multiples, which will be discussed in chapter 6) so that you will be prepared to teach these topics for understanding. Even if you plan to teach younger children, keep in mind that you will be laying the foundation for your students' later learning, therefore you must know how a topic develops beyond the grade level you teach.

2.1   Introduction to the Number Systems

A number system is a collection of numbers that fit together in a natural way. This section provides definitions of the various number systems, some interpretations and ways of representing numbers, and an overview of how the different number systems are related. What humans consider to be numbers has evolved over the course of history, and the way children learn about numbers parallels this development. When did humans first become aware of numbers? The answer to this is uncertain, but it is at least many tens of thousands of years ago, if not right from the beginning of human existence. In fact, the latter is likely, because even some animals have a primitive understanding of numbers (see [13] for a fascinating account of this as well as of the human mind's capacity to comprehend numbers).

The Counting Numbers

The simplest and most basic system of numbers is the **counting numbers**, also called the **natural numbers**.   The counting numbers are

$$1, \ 2, \ 3, \ 4, \ 5, \ldots.$$

When young children first learn to count, they simply recite the familiar sequence $1, 2, 3, \ldots$ in order without associating quantities to these numbers. When the counting numbers are used in this way, as a sequence in order, or to denote order of something in a sequence, as in

$$1^{\text{st}}, 2^{\text{nd}}, 3^{\text{rd}}, 4^{\text{th}}, 5^{\text{th}}, \ldots$$

then the counting numbers are referred to as **ordinals** or **ordinal numbers** (see Figure 2.1).

ordinal numbers:

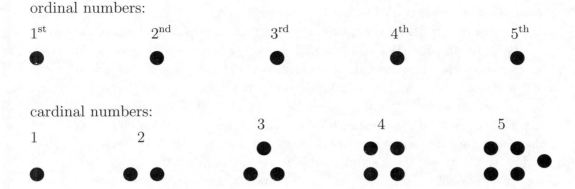

Figure 2.1: Ordinal Numbers and Cardinal Numbers

Later in their development, children learn that the counting numbers are not just a sequence to be said in order, but represent how many things are in a set. In other words, children learn that a number such as 3 is not just "what you say after 2" but can represent how many cookies are on a plate or how many balls are in a box, and so on. When counting numbers are used in this way, to represent an amount, then the counting numbers are referred to as **cardinals** or **cardinal numbers** (see Figure 2.1). To remember the distinction between *ordinals* and *cardinals* remember that *ordinals* are related to putting things in *order*.

Ordinals and cardinals are closely related to each other because to determine how many objects are in a set, we count them off one by one.

The Whole Numbers and the Integers

The **whole numbers** are the counting numbers together with zero:

$$0, \ 1, \ 2, \ 3, \ 4, \ 5, \ldots.$$

The **integers** are the counting numbers, zero (0), and the negatives of the counting numbers:

$$\ldots, \ -5, \ -4, \ -3, \ -2, \ -1, \ 0, \ 1, \ 2, \ 3, \ 4, \ 5, \ldots.$$

The **negative** of a counting number is denoted with a **minus sign**: $-$. For example, -4 is the negative of 4. The number -4 can be read *minus four* or *negative four*. When discussing the integers, the counting numbers are often referred to as **positive integers**, and their negatives as **negative integers**.

The notion of zero and of negative integers may seem natural and easy ideas to us today, but our early ancestors struggled to discover and make sense of these concepts. Although humans have probably always been acquainted with the notion of "having none" as in having no sheep or having no food to eat, the concept of 0 *as a number* was far later in coming than the notion of the counting numbers—not until sometime before 800 A.D. (see [7]). Even today, the notion of 0 is still difficult for many children to grasp. This is not too surprising because although the counting numbers can be represented nicely by sets of objects, such as in Figure 2.2, you have to show *no* objects in order to represent the number 0 in a similar fashion. But how does one *show* no objects? We can use a picture like Figure 2.3, for example.

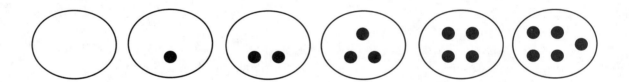

Figure 2.2: Representing the Counting Numbers

Figure 2.3: Representing the Whole Numbers

Similarly, the notion of negative numbers arose relatively late in human history. Although the ancient Babylonians may have had the concept of negative numbers in around 2000 B.C., negative numbers were not always accepted by mathematicians even as late as the 16th century A.D. ([7]). The difficulty lies in interpreting the meaning of negative numbers. Once again, starting with the representation of the counting numbers in Figure 2.2, how can negative numbers be represented in a similar fashion? This may seem perplexing at first, but in fact there is a nice interpretation of negative numbers along these lines, namely as *amounts owed*. For example, −7 can represent *owing* 7 marbles. In comparison, 7 can represent *having* 7 marbles (see Figure 2.4).

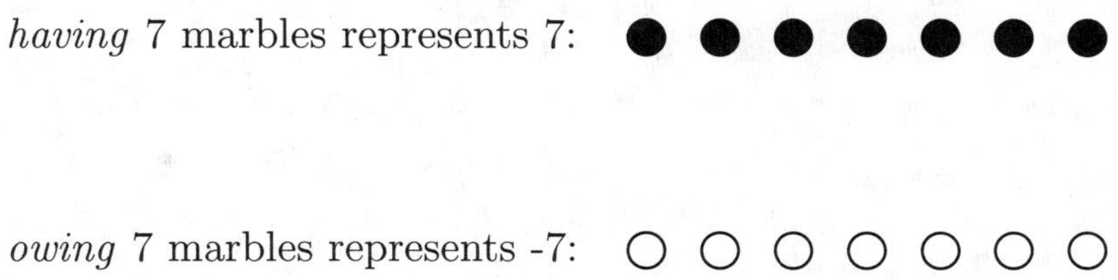

having 7 marbles represents 7: ● ● ● ● ● ● ●

owing 7 marbles represents -7: ○ ○ ○ ○ ○ ○ ○

Figure 2.4: Representing Negative Integers

A nice, although more abstract, way of representing the integers is on a number line, as in Figure 2.14. On a number line, the positive integers, 0, and the negative integers are all displayed on an equal footing. The positive integers are placed at equal spacings, heading infinitely to the right, the negative integers are placed at the same equal spacings, heading infinitely to the left, and 0 lies right between the positive integers and the negative integers.

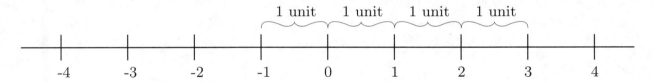

Figure 2.5: Representing Integers on a Number Line

The Rational Numbers

A **fraction** is an expression of the form

$$\frac{A}{B} \quad \text{or} \quad A/B$$

where B is not zero.

The **rational numbers** are all the numbers that can (at least theoretically) be written as a fraction

$$\frac{A}{B} \quad \text{or} \quad A/B$$

where A and B are integers and B is not zero. So, for example,

$$\frac{7}{8} \quad \text{and} \quad \frac{43}{7} \quad \text{and} \quad \frac{-2}{3}$$

are examples of rational numbers. You can remember the term *rational numbers* by remembering that *rational* has *ratio* as its root, and ratios and fractions are closely related.

A number that can be expressed as a fraction of integers is still considered to be a rational number even when the number is written in a different form. For example, we can write

$$\frac{7}{8}$$

as the decimal

$$.875$$

(by dividing 8 into 7). Therefore the number .875 is a rational number, because it can also be written as a fraction of integers (with non-zero denominator). And, we can also write

$$\frac{43}{7}$$

as the mixed number

$$6\frac{1}{7}.$$

Again, $6\frac{1}{7}$ is a rational number because it can also be represented as a fraction of integers (with non-zero denominator). These and other facts about representing rational numbers and interpreting the meaning of rational numbers will be discussed in detail in Sections 2.4 and 3.3.

It's easy to imagine how rational numbers first came into use. One can imagine a farmer filling bags with grain, the last bag perhaps only half full, or a vintner filling barrels with grape juice for wine, the last barrel only about $\frac{3}{4}$ full. Even a relatively primitive barter economy would need to use simple fractions such as $\frac{1}{2}$ and $\frac{3}{4}$ to function well.

Even very young children can develop a basic understanding of very simple fractions such as $\frac{1}{2}$, $\frac{1}{3}$, and $\frac{1}{4}$.

The Real Numbers

Before we can define the real numbers, we must define *decimals*.

A **decimal** is a written number that is in one of the following types of **decimal** forms:

$$2001$$
$$-583$$
$$37.49$$
$$-.000792$$
$$33.33333333\ldots$$
$$-127.852222222222\ldots.$$

Specifically, a decimal is a written number that is in the form

$$***\ldots***.*******\ldots$$

or

$$-***\ldots***.*******\ldots,$$

where each $*$ can be any one of the ten **digits**

$$0, 1, 2, 3, 4, 5, 6, 7, 8, 9,$$

and where there are only finitely many non-zero entries to the left of the **decimal point**, ., and either finitely or infinitely many non-zero entries to the right of the decimal point. The word *decimal* comes from the Latin word *decem* for *ten*. It is appropriate because decimals make use of the ten digits 0–9.

There are several conventions to observe when writing decimals. Leading zeros are not written in decimals. So, for example, we do not write

$$0034.7$$

but instead we write

$$34.7.$$

Also, zeros at the end of a decimal are often not written. For example, instead of writing

$$589.32000000000000\ldots$$

or

$$589.32000$$

we may write

$$589.32.$$

(Sometimes it is not proper to drop zeros on the right: when a number records a physical measurement, such as 12.0 meters, zeros to the right indicate the precision of the measurement, and so are not dropped.) When there are no non-zero entries to the right of the decimal point, the decimal point itself is usually dropped. For example, rather than writing

$$2001.0000000\ldots$$

we can simply write

$$2001$$

without any decimal point.

Another convention that is often used in writing decimals is to place commas so as to make the decimal more readable. This is especially helpful when there are many digits in the decimal. Commas are only used to the left of the decimal point. If commas are used, they are placed in front of every third digit, starting from the decimal point, going to the left. For example, we may write the decimal

$$1234567.89$$

as

$$1,234,567.89$$

in order to make it more readable.

The conventions that we use in the U.S. are not universal: in some countries the roles of the comma and the decimal point are interchanged.

When a number is written as a decimal, we sometimes refer to this decimal as the **decimal representation** or **decimal expansion** of the number, or we say that the number is written in **decimal notation**.

The **real numbers** consist of all the numbers that can, at least theoretically, be written as decimal.

So

$$1998 \quad \text{and} \quad -123.456789$$

are real numbers. The number

$$33.3333\ldots,$$

where the threes *continue forever* is another real number. The rational number

$$\frac{1}{11}$$

is also a real number, because $\frac{1}{11}$ can also be written as the decimal

$$.09090909\ldots,$$

by dividing 1 by 11 (the 09 pattern repeats forever). The square root of two,

$$\sqrt{2},$$

is yet another example of a real number because it can be written as a decimal, namely

$$1.414213562\ldots.$$

It turns out that this decimal goes on forever, without repeating. The symbol

$$\sqrt{}$$

stands for *square root*. If N is a number, then $\sqrt{N}$ stands for the positive number M such that $M \times M = N$. So, for example, $\sqrt{9} = 3$ because $3 \times 3 = 9$.

We encounter real numbers every day when we see a price such as $1.99. So when young children learn about money, they are beginning to learn about decimals and real numbers.

As with integers, the **negative** real numbers are those real numbers that have a minus sign in front when written as a decimal. The **positive** positive real numbers are those that do not have a minus sign in front when written as a decimal, with the exception of the number 0. It is understood that $-0 = 0$, and that 0 is neither positive nor negative.

As with integers, we can think of negative real numbers as represented by *owing*. For example, -35.47 can be represented by owing $35.47 on a credit card.

Relationships Between the Number Systems

Notice the following relationships:

> Every counting number is a whole number.

> Every whole number is an integer.

Every integer is a rational number because if A is an integer, then you can also write it as $\frac{A}{1}$, which shows that it is a rational number.

Every rational number is a real number because if you have a fraction $\frac{A}{B}$, you can divide B into A to get a decimal. (We will discuss why this makes sense in chapter 5.)

But none of these sets of numbers are the same:

Some real numbers are not rational numbers. For example, $\sqrt{2}$ is a real number that is not a rational number. (This is not at all obvious. It will be discussed in chapter 11.)

Some rational numbers are not integers. For example, $\frac{1}{2}$ is a rational number but it is not an integer.

Some integers are not whole numbers. For example, -1 is an integer but it is not a whole number.

0 is a whole number that is not a counting number.

Figure 2.6 shows the relationships between the kinds of numbers. All the various kinds of numbers that we will be working with—whole numbers, integers, rational numbers (fractions) and real numbers (decimals)—all these kinds of numbers are real numbers. So *the real numbers unify all the kinds of numbers.*

Exercises for Section 2.1

Answers to these exercises begin on page 31.

1. Define the the following terms:

   counting numbers

   ordinals

   cardinals

   whole numbers

   integers

   rational numbers

   real numbers

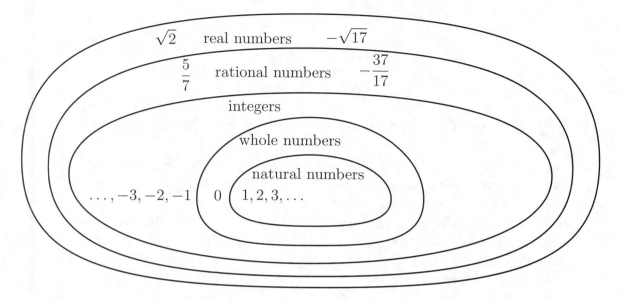

Figure 2.6: Relationships Among the Number Systems

2. Is 3 a rational number?

3. Give an example of a rational number that is not an integer.

4. Give an example of an integer that is not a whole number.

5. Is $\frac{2}{3} \times 6\frac{1}{5}$ a rational number?

Answers to Exercises for Section 2.1

1. See text.

2. Yes, 3 is a rational number because you can also write it as $\frac{3}{1}$.

3. $\frac{2}{3}$ is an example of a rational number that is not an integer.

4. -2 is an example of an integer that is not a whole number.

5. Yes, $\frac{2}{3} \times 6\frac{1}{5}$ is a rational number because you can also write it as $\frac{62}{15}$ since:

$$\frac{2}{3} \times 6\frac{1}{5} \;=\; \frac{2}{3} \times \frac{31}{5}$$

$$= \frac{2 \cdot 31}{3 \cdot 5}$$
$$= \frac{62}{15}$$

2.2   The Decimal System and Place Value

The last section stated that all the kinds of numbers we will study are also real numbers, and can be written as decimals (remember that a number such as 541, in which the decimal point has been dropped, is still considered a decimal). In this section, we will look at the meaning of a number written as a decimal, and we will consider the origins of writing numbers as decimals. Although you are certainly familiar with writing decimals, so much so that you probably take it for granted, it is actually a remarkably efficient and clever way to write numbers. It also took many thousands of years to develop. This is probably why it still takes children many years to understand what written numbers stand for.

The Origins of the Decimal System

The **decimal system** is the way of writing numbers as decimals. Sometimes this system is called the **base-ten system** for writing numbers. How did this way of writing numbers come about?

The earliest humans lived in small groups and gathered their food daily. These humans probably did not have much of a need for numbers. But eventually, some people began to live in larger groups. Then people began to specialize in their daily tasks, and to trade goods among each other. As this progressed, people developed a need for recording numbers.

For example, think back to a shepherd living thousands of years ago, long before the invention of writing. How did the shepherd keep track of his sheep every day as he led his flock from place to place? He may have made notches on a piece of wood or bone—one notch for each sheep. But if the shepherd had many sheep, and if he did not organize his notches, he might have had a hard time telling if he really had the same number of sheep on one day as on the next (see Figure 2.7). The shepherd might have come upon the idea of organizing his notches, as in Figure 2.8. This is a first step toward an efficient way of recording numbers.

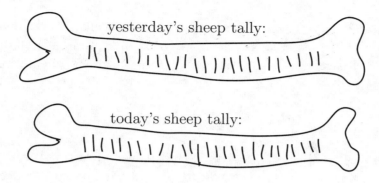

Figure 2.7: A Shepherd's Tally of Sheep

Figure 2.8: A Shepherd's Grouped Tally of Sheep

As civilization progressed, some people began to live in cities, and trade became more sophisticated. Then people needed to work with larger numbers, and so they replaced grouped tally marks, such as those in Figure 2.8, with symbols. Different cultures used different symbols. For example, you are probably familiar with the Roman numeral V which represents 5. To record 50 sheep, a person long ago might have written

$$\text{VVVVVVVVVV.}$$

But it is hard to read all those Vs, so the Romans invented new symbols: X for 10, L for 50, C for 100, and M for 1000. These are fine for representing numbers up to a few thousand, but what about representing 10,000? Once again,

$$\text{MMMMMMMMMM}$$

is difficult to read. What about 100,000, or even 1,000,000? To represent these, you'd want to invent yet more new symbols.

Eventually, these methods became too cumbersome to represent large numbers, and this led to the flexible decimal system in which we can (at least theoretically) write any number of any size without needing to invent new symbols. The key innovation of the decimal system is that rather than using new symbols to represent larger and larger numbers, the decimal system uses *place value*.

Place Value

A key feature of the decimal system is that it uses **place value** in which *the meaning of a digit in a decimal depends on its location* in the decimal. For example, the 2 in 237 stands for two *hundreds*, whereas the 2 in 14.28 stands for two *tenths*. Specifically, each position or place in a decimal corresponds to one of the following values:

Values of places in a decimal, proceeding to the left of the decimal point:

1	one
10	ten
100	hundred
1000	thousand
10000	ten-thousand
100000	hundred-thousand
1000000	million
$\vdots$	

Values of places in a decimal, proceeding to the right of the decimal point:

$\frac{1}{10}$	tenth
$\frac{1}{100}$	hundredth
$\frac{1}{1000}$	thousandth
$\frac{1}{10000}$	ten-thousdandth
$\frac{1}{100000}$	hundred-thousdandth
$\frac{1}{1000000}$	millionth
$\vdots$	

A digit in a decimal stands for that many of its place's value. For example, Figure 2.9 shows the meaning of each digit in the decimal 71294385.6023.

A decimal stands for the total amount represented by all its places taken together. For example, 3618.95 stands for

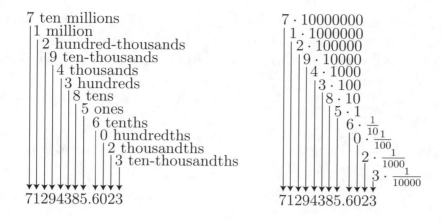

Figure 2.9: The Values That Digits in a Decimal Represent

3 thousands and

6 hundreds and

1 ten and

8 ones and

9 tenths and

5 hundredths.

We can put this in a more symbolic form by writing one of the following three expressions:

3 thousands + 6 hundreds + 1 ten + 8 ones + 9 tenths + 5 hundredths,

or

$$3 \cdot 1000 \ + \ 6 \cdot 100 \ + \ 1 \cdot 10 \ + \ 8 \cdot 1 \ + \ 9 \cdot \frac{1}{10} \ + \ 5 \cdot \frac{1}{100},$$

or

$$3000 + 600 + 10 + 8 + \frac{9}{10} + \frac{5}{100}.$$

When a decimal is re-written as one of these last three kinds of expressions, then we say that the decimal has been put into **expanded form**.   The expanded form of a decimal shows the amount represented by the decimal more explicitly than just the decimal itself because the expanded form shows the values of the places. One reason that children have difficulty learning what written numbers represent is because they must keep the values of the places in mind. Because the values of the places in a decimal are not shown explicitly, even the *interpretation* of written numbers requires a certain level of abstract thinking.

The following are some additional examples of decimals written in expanded form:

$$94,152 = 9 \cdot 10,000 \ + \ 4 \cdot 1000 \ + \ 1 \cdot 100 \ + \ 5 \cdot 10 \ + \ 2 \cdot 1,$$

$$12.438 = 1 \cdot 10 \ + \ 2 \cdot 1 \ + \ 4 \cdot \frac{1}{10} \ + \ 3 \cdot \frac{1}{100} \ + \ 8 \cdot \frac{1}{1000}.$$

Values of Places in Decimals and Powers of Ten

The decimal system for writing numbers uses place value, but how were the values of the places chosen? A key feature of the decimal system is that *the value of each place is ten times the value of the place to its immediate right.*

The values of the places in a decimal can therefore be described by repeatedly multiplying by 10:

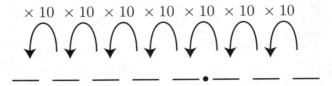

One	$1 =$	1
Ten	$10 =$	10
Hundred	$100 =$	10×10
Thousand	$1000 =$	$10 \times 10 \times 10$
Ten thousand	$10,000 =$	$10 \times 10 \times 10 \times 10$
Hundred-thousand	$100,000 =$	$10 \times 10 \times 10 \times 10 \times 10$
Million	$1,000,000 =$	$10 \times 10 \times 10 \times 10 \times 10 \times 10$

A convenient notation for writing an expression like

$$10 \times 10 \times 10 \times 10 \times 10 \times 10,$$

which is six tens multiplied together, is

$$10^6.$$

So

$$1,000,000 = 10 \times 10 \times 10 \times 10 \times 10 \times 10 = 10^6.$$

The expression 10^6 is read "ten to the sixth power" or just "ten to the sixth", and we can refer to an expression like 10^6 as a **power of ten**. The number 6 in 10^6 is called the **exponent** of 10^6. Notice the correlation between the six zeros in $1,000,000$ and the 6 in the expression 10^6.

    Similarly, the expression

power

exponent

$$10^{17}$$

stands for 17 tens multiplied together,

$$10^{17} = 10\times10\times10\times10\times10\times10\times10\times10\times10\times10\times10\times10\times10\times10\times10\times10\times10.$$

The exponent of 10^{17} is 17, and the decimal representation of 10^{17} is a 1 followed by 17 zeros:

$$10^{17} = 100,000,000,000,000,000.$$

More generally:

Ten	$10 = 10^1$
Hundred	$100 = 10 \times 10 = 10^2$
Thousand	$1000 = 10 \times 10 \times 10 = 10^3$
Ten thousand	$10,000 = 10 \times 10 \times 10 \times 10 = 10^4$
Hundred thousand	$100,000 = 10 \times 10 \times 10 \times 10 \times 10 = 10^5$
Million	$1,000,000 = 10 \times 10 \times 10 \times 10 \times 10 \times 10 = 10^6$
Ten million	$10,000,000 = 10^7$
Hundred million	$100,000,000 = 10^8$
Billion	$1,000,000,000 = 10^9$
Ten billion	$10,000,000,000 = 10^{10}$
Hundred billion	$100,000,000,000 = 10^{11}$
Trillion	$1,000,000,000,000 = 10^{12}$

And so on. Notice the pattern: the number of zeros behind the 1 (in $1,000,000$, for example) is the same as the exponent on the ten. For example, $10,000$ is written as a one followed by four zeros; it also can be written as 10^4, where the exponent on the ten is 4.

By using exponents, we can show more clearly the structure of the values of the places in decimals:

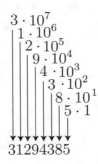

$$
\begin{array}{c}
3 \cdot 10^7 \\
1 \cdot 10^6 \\
2 \cdot 10^5 \\
9 \cdot 10^4 \\
4 \cdot 10^3 \\
3 \cdot 10^2 \\
8 \cdot 10^1 \\
5 \cdot 1
\end{array}
$$

$$31294385$$

We can also use exponents in expanded forms:

$$17,843 = 1 \cdot 10^4 \ + \ 7 \cdot 10^3 \ + \ 8 \cdot 10^2 \ + \ 4 \cdot 10^1 \ + \ 3 \cdot 1.$$

What about decimals with places to the right of the decimal point? As before, there is special notational to write one tenth, one hundredth, one thousandth, and so on, as powers of 10. This time the powers will be *negative*. Namely:

$$
\begin{array}{ccccc}
\frac{1}{10} & = & \frac{1}{10^1} & = & 10^{-1} \\
\frac{1}{100} & = & \frac{1}{10^2} & = & 10^{-2} \\
\frac{1}{1000} & = & \frac{1}{10^3} & = & 10^{-3} \\
\frac{1}{10000} & = & \frac{1}{10^4} & = & 10^{-4}
\end{array}
$$

and so on. Notice that since $1/10 = .1$ and $1/100 = .01$ and $1/1000 = .001$, and so on, the negative exponents fit the pattern of positive exponents:

$$
\begin{aligned}
10,000 &= 10^4 \\
1,000 &= 10^3 \\
100 &= 10^2 \\
10 &= 10^1 \\
1 &= 10^0 \\
.1 &= 10^{-1} \\
.01 &= 10^{-2} \\
.001 &= 10^{-3} \\
.0001 &= 10^{-4}
\end{aligned}
$$

Notice that it makes sense to *define* 10^0 to be the number one—that clearly fits with the pattern of moving the decimal point one place to the left (moving down the left hand column of numbers), and lowering the exponent on the 10 by one (moving down the column on the right).

Using the notation of exponents, we can also write 12.438 in expanded form as

$$12.438 = 1 \times 10 + 2 \times 1 + 4 \times 10^{-1} + 3 \times 10^{-2} + 8 \times 10^{-3}.$$

Class Activity 2A: Showing Powers of Ten

See the website
    `micro.magnet.fsu.edu/primer/java/scienceopticsu/powersof10/index.html`
for an amazing experience of zooming in through powers of ten.

Saying Decimals and Writing Decimals with Words

The decimal $123,456$ represents the amount

1 hundred-thousand and

2 ten-thousands and

3 thousands and

4 hundreds and

5 tens and

6 ones.

But in most circumstances, this is not how we should read this number. Instead, we should read 123, 456 as

one-hundred twenty-three thousand, four-hundred fifty-six,

separating the "thousands portion" of the number from the "hundreds portion". Similarly, when there are millions or billions involved, we read each of the "billions portion", "millions portion", "thousands portion", and "hundreds portion" separately. For example, we read

$$3, 478, 256, 149$$

as

three-billion, four-hundred seventy-eight million, two-hundred fifty-six thousand, one-hundred forty-nine.

When there are not too many digits to the right of the decimal point, we can read the decimal point as *and*. For example, we can read

$$139.7$$

as

one-hundred thirty-nine, and seven tenths.

Digits to the right of the decimal point can be read as a group. For example, we can read

$$12.37$$

as

twelve and thirty-seven hundredths,

and we can read

$$1.358$$

as

one and three-hundred fifty-eight thousdandths.

However, a number with many digits to the right of its decimal point, such as

$$5.678134$$

is best read as

five point six seven eight one three four.

There is one special number name that many children find amusing and fascinating: a *googol*. A **googol** is the name for the number whose decimal representation is a 1 followed by one hundred zeros:

$$1 \underbrace{0000000000\ldots0000000000}_{100 \text{ zeros}}.$$

A googol is so large that, according to current theories in physics, it is even larger than the number of atoms in the universe. Another very large number that has a special name is a *googolplex*. A **googolplex** is the name for the number whose decimal representation is a 1 followed by a googol zeros:

$$1 \underbrace{0000000000\ldots0000000000}_{\text{a googol zeros}}.$$

It is very difficult to comprehend such a number.

Exercises for Section 2.2 on the Decimal System and Place Value

Answers to these exercises begin on page 43.

1. What does it mean to say that the decimal system uses *place value*?

2. How are the values of adjacent places in a decimal related?

3. Write the following decimals as powers of 10:

    (a) $10,000,000,000,000$

    (b) .1

    (c) .000001

    (d) 1

    (e) 10

    (f) the number whose decimal representation is a 1 followed by 200 zeros

    (g) a googol

    (h) a googolplex

4. Write the following decimals in expanded form:

    (a) 102.04

    (b) $300,000,027$

    (c) $5,000.00200$

5. Write the following numbers in ordinary decimal notation:

    (a) One-hundred thirty-seven million

    (b) One-million thirty-seven

    (c) One-million thirty-seven thousand

6. We call 10^6 a million, we call 10^9 a billion, and we call 10^{12} a trillion. What are 10^{13} and 10^{14} called? If we call 10^{15} a thousand-trillion and 10^{18} a million-trillion, then what are 10^{19}, 10^{20}, and 10^{21} called?

7. The children in Mrs. Watson's class made chains of small paper dolls, as pictured in Figure 2.10. A chain of 5 dolls is 1 foot long. How long would the following chains of dolls be? In each case, give your answer in either feet or miles, depending on which answer is easiest to understand. Recall that 1 mile = 5280 feet.

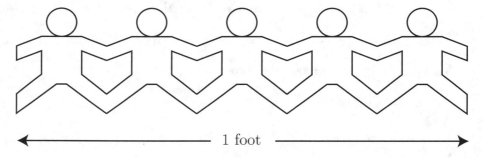

Figure 2.10: A Chain of Paper Dolls

    (a) 100 dolls

    (b) 1000 dolls

(c) 10,000 dolls

(d) 100,000 dolls

(e) one million dolls

(f) one billion dolls

Answers to Exercises for Section 2.2 on the Decimal System and Place Value

1. See text.

2. See text.

3. (a) $10,000,000,000,000 = 10^{13}$

   (b) $.1 = 10^{-1}$

   (c) $.000001 = 10^{-6}$

   (d) $1 = 10^0$

   (e) $10 = 10^1$

   (f) The number whose decimal representation is a 1 followed by 200 zeros can be written 10^{200}.

   (g) A googol can be written 10^{100}.

   (h) A googolplex can be written 10^{googol} or $10^{(10^{100})}$.

4. (a) $1 \cdot 100 + 2 \cdot 1 + 4 \cdot \frac{1}{100}$ or
       $1 \cdot 10^2 + 2 \cdot 10^0 + 4 \cdot 10^{-2}$

   (b) $3 \cdot 100,000,000 + 2 \cdot 10 + 7 \cdot 1$ or
       $3 \cdot 10^8 + 2 \cdot 10^1 + 7 \cdot 10^0$

   (c) $5 \cdot 1000 + 2 \cdot \frac{1}{1000}$ or
       $5 \cdot 10^3 + 2 \cdot 10^{-3}$

5. (a) $137,000,000$

   (b) $1,000,037$

   (c) $1,037,000$

6. Since 10^{12} is a trillion, therefore 10^{13} is ten-trillion and 10^{14} is a hundred-trillion. Since 10^{18} is a million-trillion, therefore 10^{19} is ten-million-trillion, 10^{20} is a hundred-million-trillion, and 10^{21} is a billion-trillion.

7. (a) Because 5 dolls are 1 foot long, therefore 10 dolls are 2 feet long, and so 100 dolls are 10 times as long, namely 20 feet long.

   (b) 1000 dolls are 10 times as long as 100 dolls. Because 100 dolls are 20 feet long, therefore 1000 dolls are 200 feet long.

   (c) 10,000 dolls are 10 times as long as 1000 dolls. Because 1000 dolls are 200 feet long, therefore 10,000 dolls are 2000 feet long.

   (d) 100,000 dolls are 10 times as long as 10,000 dolls. Because 10,000 dolls are 2000 feet long, therefore 100,000 dolls are 20,000 feet long. One mile is 5280 feet, so 20,000 feet is $20,000 \div 5280 = 3.8$ miles. So 100,000 dolls are nearly 4 miles long.

   (e) One million dolls are 10 times as long as 100,000 dolls. Because 100,000 dolls are 3.8 miles long, therefore one million dolls are 38 miles long.

   (f) One billion dolls are 1000 times as long as one million dolls. Because one million dolls are 38 miles long, therefore one billion dolls are 38,000 miles long. The circumference of the earth is about 24,000 miles, so one billion dolls would wrap around the earth more than $1\frac{1}{2}$ times!

Problems for Section 2.2 on the Decimal System and Place Value

1. Write 100 zeros on a piece of paper and time how long it takes you. Based on how long it took you to write 100 zeros, approximately how long would it take you to write 1,000 zeros? 10,000 zeros? 100,000 zeros? one million zeros? one billion zeros? one trillion zeros? In each case, give your answer either in minutes, hours, days, or years, depending on which answer is easiest to understand. Explain your answers.

2. Explain clearly why it would not be possible for anyone to write a googol zeros. Use this to explain why no person could ever write a googolplex as a decimal.

3. The students in Ms. Caven's class have a large poster showing a million dots. Now the students would really like to see a billion of something. Think of at least two different ways that you might attempt to show a billion of something and discuss whether or not your methods would be feasible. Be specific and back up your explanations with calculations.

2.3   Representing Decimals

We will study two ways to represent decimals. One is with physical objects. Another is with number lines.

Representing Decimals with Physical Objects

Writing a decimal in its expanded form can help you understand the meaning of the decimal, but physical objects, while cumbersome, may be even more helpful. Appropriate physical objects can be especially helpful for developing a feel for place value.

Using Money to Represent Decimals

A familiar way to represent decimals is with money. For example,

$$12.34 = 1 \cdot 10 \ + \ 2 \cdot 1 \ + \ 3 \cdot \frac{1}{10} \ + \ 4 \cdot \frac{1}{100}$$

can be represented by:

1 ten-dollar bill,

2 one-dollar bills,

3 dimes, and

4 pennies.

Notice that this same set of bills and coins can also represent

$$1234 = 1 \cdot 1000 \ + \ 2 \cdot 100 \ + \ 3 \cdot 10 \ + \ 4 \cdot 1,$$

when we consider a penny as 1 *cent* rather than $\frac{1}{100}$ of a *dollar* (see the top of Figure 2.13).

Even though money is familiar, one disadvantage of using money to represent decimals is that it does not show that the value of each place in a decimal is 10 times the value of the place to the right. For example, we can't tell just by looking at a dollar bill that its value is ten times the value of a dime. Although every adult knows this, it is not an apparent feature from the way a dollar bill and a dime look.

Using Bundled Objects to Represent Decimals

One way to physically represent decimals so that it is apparent that each place's value is ten times the value of the place to the right is to use bundled objects. Toothpicks are convenient since they are small, inexpensive, and readily available (coffee stirers also work well if you can get enough of them). To represent the various place's values, first make bundles of 10 toothpicks and band them together, or place 10 toothpicks in a small plastic sandwich bag. Then make bundles of 100 toothpicks by banding 10 bundles of 10 (you can put a rubber band around 10 bags that each contain 10 toothpicks). If you have enough time and toothpicks, you can make a bundle of 1000 toothpicks by banding 10 bundles of 100 (which themselves are 10 bundles of 10). See Figure 2.11. Even though it may seem tedious, *it is important to do this successive bundling because this is how you demonstrate that each place's value is 10 times the value of the place to the right.* If you just put 100 loose toothpicks in a plastic bag, for instance, then you don't show the structure of the hundreds place as 10×10, i.e., 10 groups of 10.

To show the decimal 1234 with toothpicks, use:

1 bundle of 1000 (which is 10 bundles of 100 — each of which is 10 bundles of 10),

2 bundles of 100 (each of which is is 10 bundles of 10),

3 bundles of 10 toothpicks, and

4 individual toothpicks,

as in Figure 2.13. However, the same configuration of bundles of toothpicks can also represent

$$123.4 = 1 \cdot 100 \; + \; 2 \cdot 10 \; + \; 3 \cdot 1 \; + \; 4 \cdot \frac{1}{10},$$

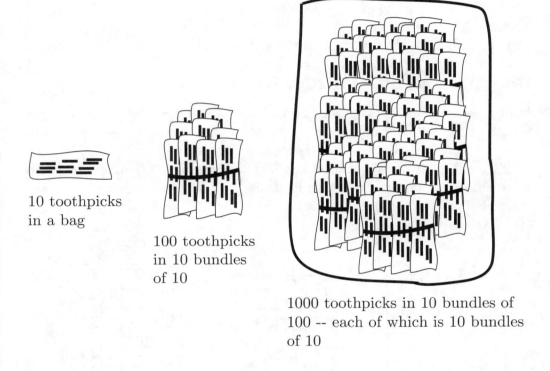

10 toothpicks
in a bag

100 toothpicks
in 10 bundles
of 10

1000 toothpicks in 10 bundles of
100 -- each of which is 10 bundles
of 10

Figure 2.11: Bundled Toothpicks for Representing Values of Places

by simply interpreting 1 toothpick as $\frac{1}{10}$, so that a bag of 10 toothpicks must represent 1, a bundle of 10 bags of 10 toothpicks must represent 10, and a bundle of 10 bundles of 100 (which itself is 10 bundles of 10) represents 100.

Similarly, the same configuration of bundled toothpicks can represent

$$12.34, \quad 1.234, \quad .1234, \quad .01234, \quad .001234, \quad .0001234,$$

and so on, as well as

$$12340, \quad 123400, \quad 1234000, \quad 12340000,$$

and so on, depending on what one toothpick represents.

Using Base Ten Blocks to Represent Decimals

Many teachers have sets of **base ten blocks** for representing decimals. Base ten blocks are sets of wooden (or plastic) blocks of several types: small cubes; "longs," that look like 10 small cubes glued together in a row; "flats," that look like 10 longs glued together to make a square of 100 small cubes; and a large cube that is constructed to look like 10 flats glued together to make a cube of 1000 small cubes (see Figure 2.12).

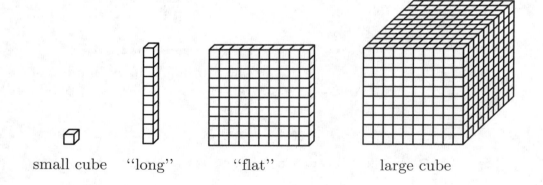

small cube   "long"   "flat"   large cube

Figure 2.12: Base Ten Blocks

To represent

$$1234 = 1 \cdot 1000 + 2 \cdot 100 + 3 \cdot 10 + 4 \cdot 1$$

with base ten blocks you can use:

1 large cube (representing 1000 small cubes),

2 flats (each representing 100 small cubes),

3 longs (each representing 10 small cubes), and

4 small cubes,

as shown in the middle of Figure 2.13. As before, this same configuration of base ten blocks can represent other decimals by changing the interpretation of 1 small cube. For example, it can represent the decimal .1234 by considering 1 small block to be $\frac{1}{10,000}$, so that therefore a long represents $\frac{1}{1,000}$, a flat represents $\frac{1}{100}$, and a large cube represents $\frac{1}{10}$.

One disadvantage of using base ten blocks is that it is not always apparent to students that for successive blocks, the larger one is made of 10 times as many small cubes as the smaller one.

Class Activity 2B: Representing Decimals with Physical Objects

Representing Decimals on Number Lines

Bundled toothpicks show nicely that the value of a place in a decimal is 10 times the value of the place to its right. Starting at any place in a decimal, you can let 1 toothpick stand for the value of that place. Moving to the *left* in the decimal, the values of subsequent places are shown by repeatedly bundling bundles. As long as you have enough toothpicks, you can keep going. But what about starting at a place in a decimal and moving to the *right*? A good way to show the decreasing values of places as one moves to the right in a decimal is with *number lines*.

A **number line** is a line on which one location has been chosen as 0, and another location, to the right of 0, has been chosen as 1. Number lines stretch infinitely far in both directions, although in practice, only a small portion of a number line can be shown (and that portion may or may not include 0 and 1). The distance from 0 to 1 is called a **unit**, and the choice of a unit is called the **scale** of the number line. Once a choice for 0 and 1 have been made, *every decimal is represented by a location on the number line, and every location on the number line corresponds to a decimal.*

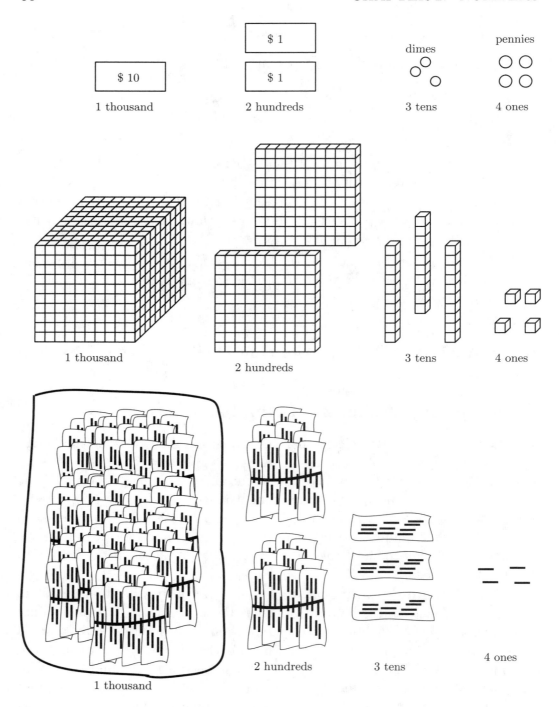

Figure 2.13: Representing $1234 = 1 \cdot 1000 + 2 \cdot 100 + 3 \cdot 10 + 4 \cdot 1$

Decimals "Fill In" the Number Line

Decimals "fill in" the locations on the number line between the integers. You can think of plotting decimals on the number line in successive stages. First, the integers are placed on a number line so that consecutive integers are one unit apart. The positive integers are to the right of 0, and the negative integers are to the left of zero (see Figure 2.14).

Figure 2.14: Representing Integers on a Number Line

Next, the decimals that have entries in the tenths place, but no smaller place, are spaced equally between the integers, breaking each interval between consecutive integers into 10 smaller intervals, each $\frac{1}{10}$ unit long. See the top two number lines shown in Figures 2.15 and 2.16. Notice that although the *interval* between consecutive integers is broken into 10 *intervals*, there are only 9 *"tick marks"* for decimals in the interval, one for each of the 9 non-zero entries 1–9 that go in the tenths places.

At the next stage, the decimals that have entries in the hundredths place, but no smaller place, are spaced equally between the previously plotted decimals, breaking each interval between previously plotted consecutive decimals into 10 smaller intervals, each $\frac{1}{100}$ unit long. See the number lines in the middle of Figures 2.15 and 2.16. Notice that the number lines shown in Figures 2.15 and 2.16 all have different scales, so as to be able to show clearly the locations of the decimals. However, these locations also exist in the first number line at the top.

Imagine this process of repeatedly breaking intervals into 10 smaller intervals continuing forever. Then all decimals, including the ones that have infinitely many non-zero entries to the right of the decimal point, will have a location on the number line. This last statement, that even decimals with infinitely many non-zero entries to the right of the decimal point have a location on the number line, may seem murky; in fact, it is a very subtle point, whose details were not fully worked out by mathematicians until the late 19th century.

The idea of "filling in" locations on the number line, starting from the

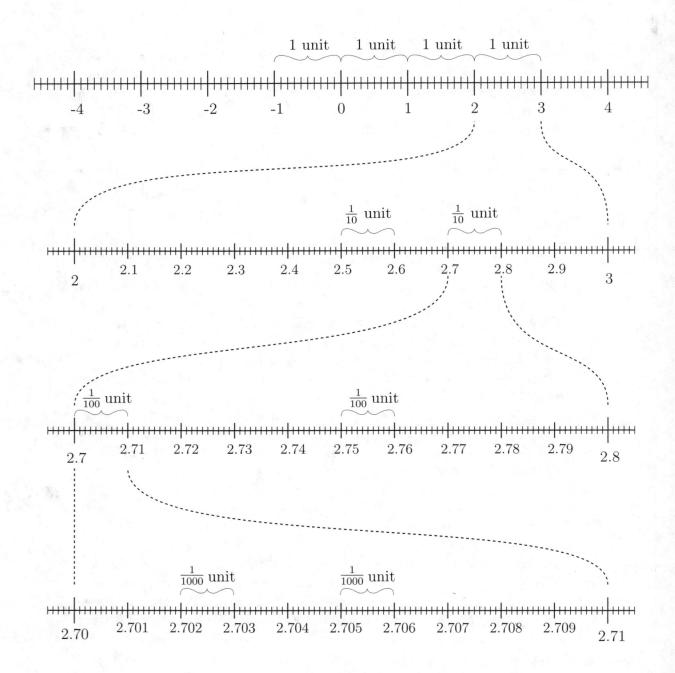

Figure 2.15: Decimals Fill In Number Lines

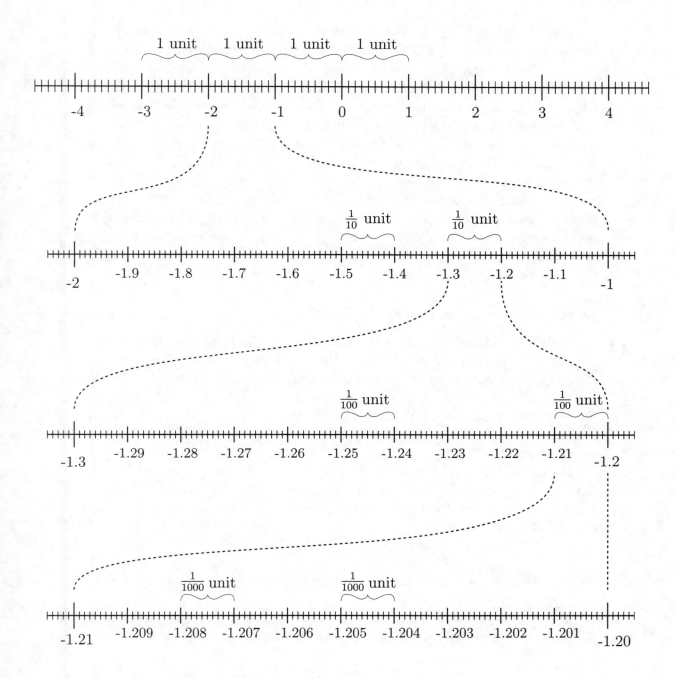

Figure 2.16: Negative Decimals Fill In Number Lines

integers, is like starting with the notion of *dollars* as a denomination and deciding that a smaller denomination is needed, therefore creating *dimes*. 10 dimes make a dollar and a dime is worth $\frac{1}{10}$ of a dollar, just as the intervals on a number line between consecutive integers are broken into 10 intervals, each $\frac{1}{10}$ of a unit long. If dimes don't make a small enough denomination, *pennies* are created. 10 pennies make a dime and a penny is worth $\frac{1}{100}$ of a dollar, just as the intervals between consecutive decimals with no entries to the right of the tenths place are broken into 10 smaller intervals, each $\frac{1}{100}$ of a unit long. If we felt that we needed a smaller denomination than the penny, we would create a new coin. In this case, 10 of the new coins would make a penny. In theory, we could keep going, creating smaller and smaller denominations of money, just as the values of places in decimals get smaller and smaller as one moves to the right, and just as we can repeatedly break intervals on a number line into 10 smaller intervals.

Decimals Represent Distances from 0

Another way to think about where a decimal should be plotted on a number line is to think in terms of distances from 0, and to consider the expanded forms of decimals. A positive decimal N is located to the right of 0, at a distance of N units away from 0. A negative decimal $-N$ is located to the left of 0, at a distance of N units away from 0. From this point of view, a number line is like an infinite ruler, except that number lines also show negative numbers.

For example, 3 is located to the right of 0 at a distance of 3 units from 0. The negative decimal -2 is located to the left of 0 at a distance of 2 units from 0. To plot a decimal that is not an integer, such as 1.3, we can use the expanded form. Since

$$1.3 = 1 \cdot 1 \ + \ 3 \cdot \frac{1}{10},$$

locate this decimal 1 unit and an additional 3 tenths of a unit to the right of zero (see Figure 2.17).

To plot a negative decimal, such as -2.41, we can work with the expanded form of the decimal obtained by dropping the minus sign, 2.41.

$$2.41 = 2 \cdot 1 \ + \ 4 \cdot \frac{1}{10} \ + \ 1 \cdot \frac{1}{100}.$$

Therefore -2.41 is located to the left of 0 at a distance of 2 units plus 4 tenths of a unit plus 1 hundredth of a unit away from 0. So to locate this decimal on

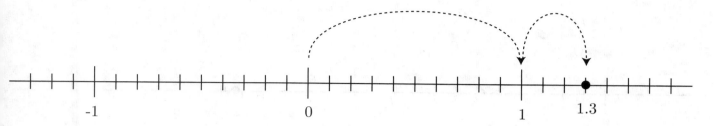

Figure 2.17: Plotting $1.3 = 1 \cdot 1 + 3 \cdot \frac{1}{10}$

the number line you start at 0, move 2 units to the left, then move another 4 tenths of a unit *to the left*, and finally move yet another 1 hundredth of a unit, *still moving to the left* (see Figure 2.18. In this way the decimal -2.41 is plotted a full 2.41 units away from 0 (if you moved to the right in the second or third step, the decimal would be plotted less than 2.41 units away from 0).

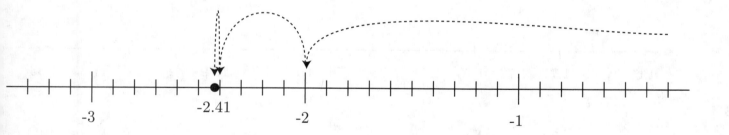

Figure 2.18: Plotting -2.41

Using Number Lines of Different Scales

Where you plot a decimal on a number line depends on the scale of the number line. For example, Figure 2.19 shows the location of 1.738 on number lines of different scales.

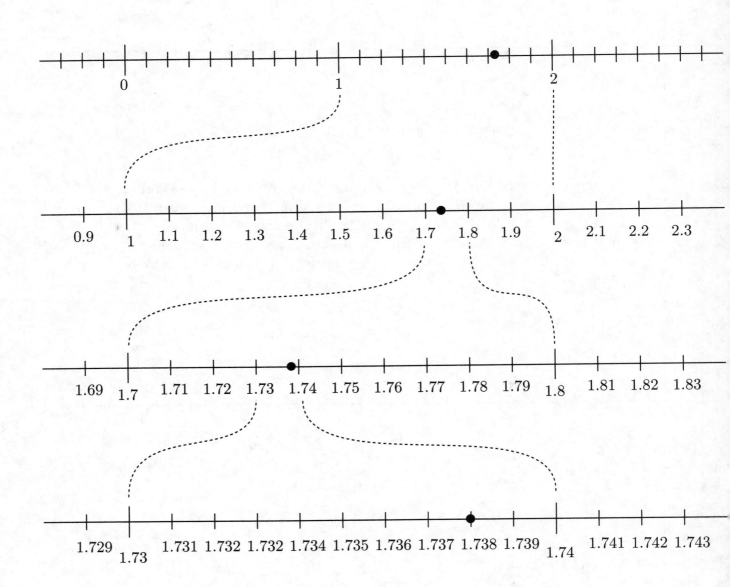

Figure 2.19: Plotting 1.738 on Number Lines With Different Scales

Decimals With Infinitely Many Non-zero Entries

Some decimals extend infinitely far to the right. For example, it turns out that the the decimal representation of $\sqrt{2}$,

$$\sqrt{2} = 1.41421356237\ldots,$$

goes on forever, never ending in zeros. *Every* decimal, even a decimal that extends infinitely to the right, has a definite location on a number line. However, such a decimal will never fall *exactly* on a tick mark, no matter what the scale of the number line. Figure 2.20 shows where $\sqrt{2}$ is located on number lines of various scales.

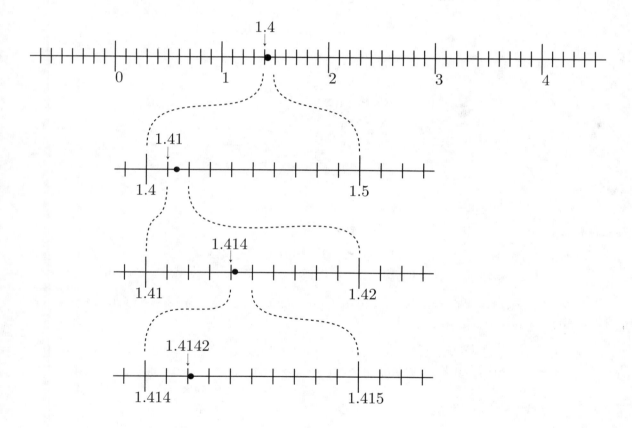

Figure 2.20: Plotting $\sqrt{2} = 1.41421356237\ldots$

Class Activity 2C: Decimals on Number Lines

Exercises for Section 2.3 on Representing Decimals

Answers to these exercises begin on page 60.

1. Bundles of toothpicks and money can both be used to represent decimals. In what way are bundles of toothpicks better at representing decimals than money is?

2. Describe a way to represent .0278 with bundles of toothpicks. In this case, what does one toothpick represent?

3. List at least three different decimals that the bundled toothpicks in Figure 2.21 could represent. In each case, state the value of one toothpick.

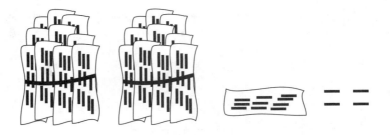

Figure 2.21: Which Decimals Can This Represent?

4. Give examples of decimals that cannot be represented with bundles of toothpicks—even if you had as many toothpicks as you wanted.

5. Label the "tick marks" on the number lines in Figure 2.22 with appropriate decimals.

6. Label the tick marks on the number line in Figure 2.23 so that the decimals 3.482179 and 3.482635 can both be plotted visibly and distinctly (they do not necessarily have to fall on tick marks). Plot the two decimals.

7. Label the tick marks on the number line in Figure 2.23 so that the decimals −7.65 and −3 can both be plotted visibly and distinctly (they do not necessarily have to fall on tick marks). Plot the two decimals.

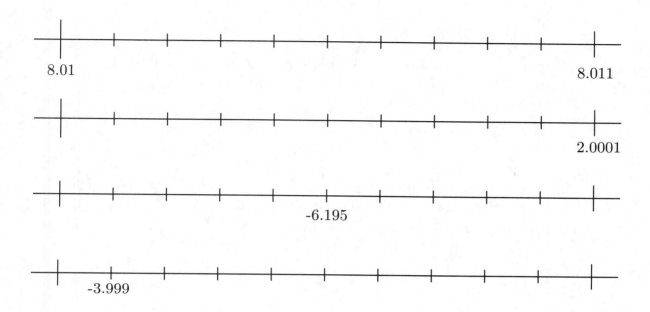

Figure 2.22: Label the Tick Marks

Figure 2.23: A Number Line

Answers to Exercises for Section 2.3 on Representing Decimals

1. See text.

2. Represent .0278 as 2 bundles of 100 toothpicks (each of which is 10 bundles of 10), 7 bundles of 10 toothpicks, and 8 individual toothpicks. In this case, each individual toothpick must represent $\frac{1}{10,000}$, since the expanded form of .0278 is

$$2 \cdot \frac{1}{100} + 7 \cdot \frac{1}{1,000} + 8 \cdot \frac{1}{10,000}.$$

3. If one toothpick represents 1, then Figure 2.21 represents 214. The table below shows several other possibilities.

If 1 toothpick represents:	then Figure 2.21 represents:
100	21,400
10	2140
1	214
$\frac{1}{10}$	21.4
$\frac{1}{100}$	2.14
$\frac{1}{1000}$	.214

4. Realistically we'd be hard pressed to represent numbers with more than 4 non-zero digits with toothpicks. Even 999 would be pretty awful to have to do. But there are some numbers whose expanded form *can't* be represented by toothpicks (in the manner described in the text)—even if you had as many toothpicks as you wanted. For example, consider

$$.333333333\ldots,$$

where the 3s go on forever. Which place's value would you pick to be represented by 1 toothpick? If you did pick such a place, you'd have to represent the values of the places to the right by tenths of a toothpick, hundredths of a toothpick, thousandths of a toothpick, and so on *forever* in order to show the expanded form of this number.

As an aside, here's something amazing: we *can* represent .333333333... with toothpicks, but in a different way (not by bundling so as to show the places in the decimal). Namely, we can represent .333333333... by

$\frac{1}{3}$ of a toothpick, because it so happens that $\frac{1}{3} = .333333333\ldots$ (which you can see by dividing 1 by 3).

5. See Figure 2.24.

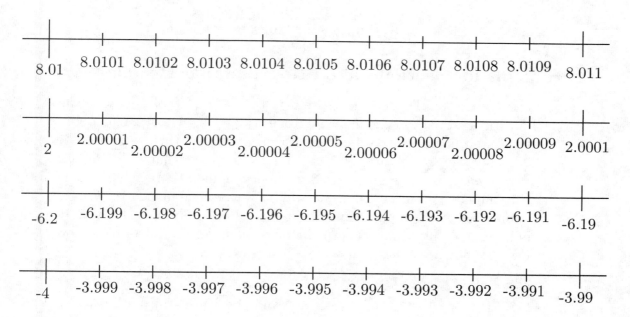

Figure 2.24: Labeled Number Lines

6. See Figure 2.25.

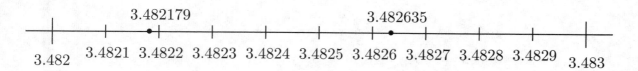

Figure 2.25: Choosing an Appropriate Scale

7. See Figure 2.26.

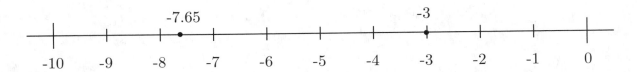

Figure 2.26: Choosing an Appropriate Scale

Problems for Section 2.3 on Representing Decimals

1. One feature of the decimal system is that every whole number can be expressed as a decimal in only one way without a decimal point. If we used a different system for writing numbers, this might not be true. Let's make up another way to write numbers that we'll call the *money numeration system*. In the money numeration system, there will generally be more than one way to write a number. To write a number using the money numeration system, simply record a way to make that many cents out of coins and bills. Several examples show how the money numeration system works:

   $23 = 2(\text{dime}) + 3(\text{cent})$, also:

   $23 = 1(\text{dime}) + 2(\text{nickel}) + 3(\text{cent})$ (and there are still more ways to write 23),

   $5,175 = 5(\text{ten dollar}) + 1(\text{dollar}) + 3(\text{quarter})$.

   (a) Write $2,742$ (cents) in the money numeration system in at least 3 different ways.

   (b) Write 27 (cents) in the money numeration system in all possible ways. Explain how you know you've found all the ways. Is it surprising how many ways there are?

   (c) Which numbers can only be written in one way in the money numeration system?

   (d) Why is it convenient to have a system of money where most amounts can be made in many ways? Why is it convenient to have a system for writing numbers (the decimal system) where every whole number can be written only in one way (without a decimal point)?

2. Explain why the bagged and loose toothpicks pictured in Figure 2.27 do not represent a decimal in the manner described in the text. Describe how to alter the appearance of these bagged and loose toothpicks so that the same total number of toothpicks *does* represent a decimal.

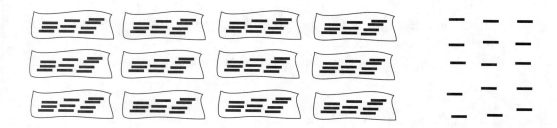

Figure 2.27: Why Do These Not Represent a Decimal?

3. List three different ways that the left-most tick mark in Figure 2.28 can legitimately be labeled. In each case, show how to label the smaller tick marks as well.

4.3

Figure 2.28: How to Label the Tick Marks?

4. Jerome says that the unlabeled tick mark on the number line in Figure 2.29 should be 7.10. Why might Jerome think this? Explain to Jerome why he is not correct. Describe how you could use physical objects to help Jerome determine the correct answer.

2.4 The Meaning of Fractions

The last two sections discussed how to interpret and represent decimals. This section will discuss the meaning of (positive) fractions, some of the common

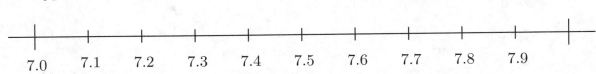

Figure 2.29: How to Label the Tick Mark on the Right?

difficulties in understanding the meaning of fractions, and how to use simple pictures to represent fractions.

First, consider how fractions are used in ordinary situations:

$\frac{1}{8}$ of a pizza,

$\frac{2}{3}$ of the houses in the neighborhood,

$\frac{9}{10}$ of the profit,

$\frac{5}{4}$ of a cup of water.

$\frac{4}{100}$ of \$20,000.

All these examples use the word *of*, and all the fractions above are *of* some object, collection of objects, or quantity.

If A and B are whole numbers, and B is not zero, then the **fraction**

$$\frac{A}{B}$$

of an object, a collection or a quantity is the amount formed by A parts (or A copies of parts) when the object, collection or quantity is divided into B equal parts.

If $\frac{A}{B}$ is a fraction, then A is called the **numerator** and B is called the **denominator**. These names make sense because the word *numerator* comes from *number* and the *numerator* tells you the *number* of parts. The word *denominator* comes from *name* and is related to *denomination*, which tells what type something is. For example, in money, the denomination of a bill tells you what type of bill it is, in other words, what value it has. In religion, a denomination is a group with the same type of religious belief and practice. The denominator of a fraction tells you what type of parts are being used. These parts can be halves, thirds, fourths, fifths, and so on. So a fraction tells you how many of what type of parts.

$$\frac{\text{A}}{\text{B}}$$

$\text{A} \quad \leftarrow \quad$ numerator: the *number* of parts

$\text{B} \quad \leftarrow \quad$ denominator: the *type* of the parts

Figure 2.30 shows some simple pictures representing fractions of objects or collections of objects. In each case, the fraction $\frac{A}{B}$ is represented by shading A parts when the whole object or collection of objects has been divided into B equal parts. In the pizza example, $\frac{1}{8}$ of the pizza is shown within the whole pizza. Likewise in the houses example: $\frac{2}{3}$ of the houses are shown shaded within the whole collection of houses. However, since $\frac{5}{4}$ cup of water is more than a whole cup of water, it is shown separate from the whole cup of water.

Class Activity 2D: Fractions Of Objects

A Fraction is Associated with a Whole

Above, we considered fractions of objects, collections of objects, or quantities. Notice the crucial word *of* in these examples:

$\frac{2}{5}$ *of* the land;

$\frac{2}{3}$ *of* the cars on the road;

$\frac{1}{10}$ *of* the water in a lake.

Fractions are defined *in relation to a whole*, and this whole can be just one object, or it can be a collection of objects (such as 24 houses). The whole can also be a quantity, such as a quantity of water, or it can simply be the number 1.

When working with fractions, always keep in mind that there is an associated whole—whether or not that whole has been made explicit or drawn to your attention. Many mistakes and misconceptions of students from elementary school through college can be corrected by paying attention to the whole associated to a fraction, in other words, what the fraction is *"of."*

One elementary school teacher, Patti Huberty, a mathematics specialist, begins her fraction lessons by asking children to find fractions of numbers, such as "what is half of 10?". Mrs. Huberty finds that this helps children focus on the associated whole.

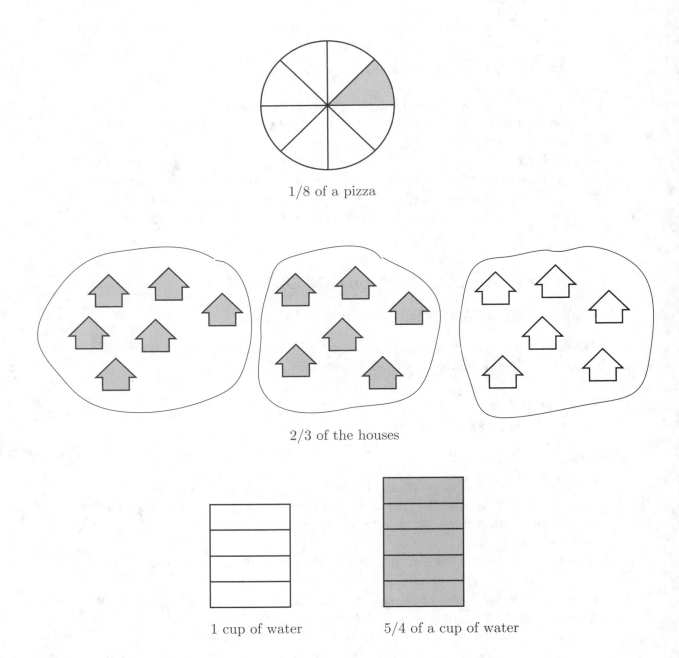

Figure 2.30: Fractions of Objects, Collections, And Quantities

Class Activity 2E: The Whole Associated to a Fraction

Equal Parts

Class Activity 2F: Is the Meaning of Equal Parts Always Clear?

If you did the previous class activity, then you saw that the meaning of *equal parts* is not always completely clear. If you want to divide a collection of toys into 4 equal parts, you might not simply put the same number of toys in each part: all the toys may not be considered equal, some may be more desirable than others. If you are asked to divide a shape, such as the one in Figure 2.31, into 3 equal parts, you might not know whether the parts should be equal in size *and shape* or if they only need to be the same size.

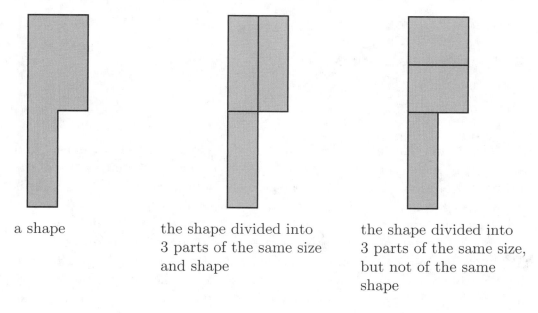

a shape   the shape divided into   the shape divided into
               3 parts of the same size   3 parts of the same size,
               and shape   but not of the same
                                               shape

Figure 2.31: What Does *Equal Parts* Mean?

In realistic situations, it is seldom a problem to determine what constitutes *equal parts*. In most cases, equal parts will be defined by dollar value, by length, by area, by volume, or by number. To divide the items in an estate into equal parts, you would probably make the parts of equal dollar value. For rope, parts of the same length constitute equal parts. For land, parts of the same area would generally constitute equal parts. For a pile of gravel, parts of the same volume constitute equal parts. On the other hand, when talking about $\frac{2}{3}$ of the stamps in a collection, or $\frac{2}{3}$ of the cars on the road, or $\frac{2}{3}$ of the people in the room, we treat each stamp, car, or person as equal, even though some stamps may be more valuable than others, some cars may be more valuable or bigger than others, and some people are bigger than others.

Understanding Improper Fractions

A fraction of whole numbers $\frac{A}{B}$ (where B is not zero) is called a **proper fraction** if the numerator A is smaller than the denominator B. Otherwise, it's called an **improper fraction**. For example,

$$\frac{2}{5}, \quad \frac{3}{8}, \quad \frac{15}{16}$$

are proper fractions, whereas

$$\frac{6}{5}, \quad \frac{11}{8}, \quad \frac{27}{16}$$

are improper fractions.

Many students at all levels, from elementary school through college, find it difficult to make sense of improper fractions. How do we make sense of the notion of $\frac{3}{2}$ of a candy bar? According to the definition, $\frac{3}{2}$ of a candy bar is the amount formed by 3 parts (or 3 copies of parts) when the candy bar is divided into 2 equal parts. In order to conceptualize $\frac{3}{2}$ of a candy bar one must be able to conceive of a third part that is a *copy* of the two equal parts that make up the whole candy bar. One must therefore be able to think of the candy bar and its parts as *replicable*. This is like being able to see 5 candy bars as 5 copies of 1 candy bar. So, two ideas are needed to make sense of $\frac{3}{2}$ of a candy bar: the idea that half of a candy bar is an entity in its own right, and the idea that half a candy bar can be replicated 3 times, even though only 2 of those copies are present in one full candy bar.

In some realistic situations, where an object is unique, an improper fraction of that object may not make sense. This probably contributes to the difficulty of making sense of improper fractions. If Mama bakes one special pie that she's never baked before and may never bake again, does it really make sense to talk about $\frac{9}{8}$ of Mama's pie? On the other hand, if we go to a pie shop where there are many identical cherry pies, the idea of $\frac{9}{8}$ of a cherry pie does make sense. When it is easy to imagine many identical copies of an object, or large amounts of a quantity, such as water, then it becomes easier to make sense of improper fractions. The idea of $\frac{5}{4}$ of a cup of water may be easier to understand than $\frac{5}{4}$ of a pie because we all experience a nearly limitless supply of water coming from our taps, and one cup of water is virtually identical to any other cup of water.

Class Activity 2G: Making Sense of Improper Fractions

Exercises for Section 2.4 on the Meaning of Fractions

1. Show $\frac{1}{8}$ of the combined amount in the 3 pies in Figure 2.32 and explain why your answer is correct.

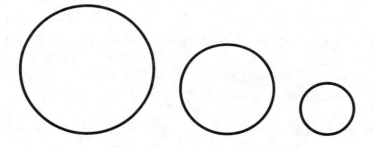

Figure 2.32: Three Pies

2. This rectangle of x's is $\frac{3}{4}$ of another (original) rectangle of x's.

$$
\begin{array}{cccc}
x & x & x & x \\
x & x & x & x \\
x & x & x & x \\
x & x & x & x \\
x & x & x & x \\
x & x & x & x \\
\end{array}
$$

Show the original rectangle.

3. This rectangle of x's is $\frac{8}{3}$ of another (original) rectangle of x's.

$$x \quad x \quad x \quad x \quad x \quad x \quad x \quad x \quad x \quad x \quad x \quad x \quad x \quad x \quad x \quad x$$
$$x \quad x \quad x \quad x \quad x \quad x \quad x \quad x \quad x \quad x \quad x \quad x \quad x \quad x \quad x \quad x$$
$$x \quad x \quad x \quad x \quad x \quad x \quad x \quad x \quad x \quad x \quad x \quad x \quad x \quad x \quad x \quad x$$

Show the original rectangle.

4. Mr. Smith owns $\frac{3}{4}$ of a rectangular plot of land that had belonged to Mr. Henley. One of Mr. Henley's heirs, Hank, is to get $\frac{1}{4}$ of Mr. Henley's original land. Hank will receive this land from Mr. Smith. What fraction of Mr. Smith's land should Hank get? Draw a picture to help you solve this problem. Explain your answer clearly. For each fraction in this problem, and in your solution, describe the *whole* that this fraction is associated with.

5. If one serving of juice gives you $\frac{3}{2}$ of your daily value of vitamin C, how much of your daily value of vitamin C will you get in $\frac{2}{3}$ of a serving of juice? Draw a picture to help you solve this problem. Explain your answer clearly. For each fraction in this problem, and in your solution, describe the *whole* that this fraction is associated with.

6. Show how to divide the shaded shape in Figure 2.33 into 4 equal parts in two different ways: one where all parts are the same size and shape, and one where all parts are the same size (area-wise) but some parts are not the same shape.

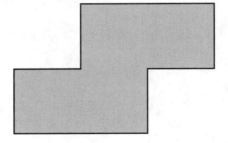

Figure 2.33: Divide Into 4 Equal Parts

Answers to Exercises for Section 2.4 on the Meaning of Fractions

1. See Figure 2.34. If you shade $\frac{1}{8}$ of *each pie individually*, then collectively, the shaded parts are $\frac{1}{8}$ of the 3 pies comined. This is because 8 of these combined parts make up the whole three pies. If you had 8 people at a party and 3 pies, you'd probably want everybody to get an equal share of all 3 pies, i.e., everybody should get $\frac{1}{8}$ of all 3 pies. How could you do this? Divide each pie into 8 equal pieces and give everybody 1 piece of each pie.

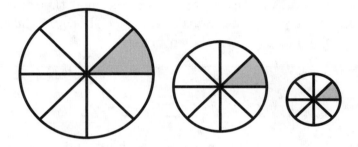

Figure 2.34: $\frac{1}{8}$ Of Three Pies

2. Because the rectangle in the exercise is $\frac{3}{4}$ of another rectangle, it must consist of 3 parts of the other rectangle, when the other rectangle is divided into 4 equal parts. So the rectangle shown in the exercise must consist of 3 equal parts, as shown below:

x	x	x	x
x	x	x	x
x	x	x	x
x	x	x	x
x	x	x	x
x	x	x	x

The original rectangle that we are looking for must consist of 4 of these parts:

$$
\begin{array}{cccc}
x & x & x & x \\
x & x & x & x \\
x & x & x & x \\
x & x & x & x \\
x & x & x & x \\
x & x & x & x \\
x & x & x & x \\
x & x & x & x \\
\end{array}
$$

3. Because the rectangle in the exercise is $\frac{8}{3}$ of another rectangle, it must consiste of 8 parts of the other rectangle, when the other rectangle is divided into 3 equal parts. So the rectangle shown in the exercise must consist of 8 equal parts, as shown below:

$$
\begin{array}{|cc|cc|cc|cc|cc|cc|cc|cc|}
\hline
x & x & x & x & x & x & x & x & x & x & x & x & x & x & x & x \\
x & x & x & x & x & x & x & x & x & x & x & x & x & x & x & x \\
x & x & x & x & x & x & x & x & x & x & x & x & x & x & x & x \\
\hline
\end{array}
$$

The orginal rectangle that we are looking for must consist of 3 of these parts:

$$
\begin{array}{cccccc}
x & x & x & x & x & x \\
x & x & x & x & x & x \\
x & x & x & x & x & x \\
\end{array}
$$

4. As Figure 2.35 shows, Mr. Smith's land is made up of 3 of the parts when Mr. Hensley's land is divided into 4 equal parts. 1 of those 3 parts is to go to Mr. Hensley's heir. Therefore Mr. Hensley's heir should receive $\frac{1}{3}$ of Mr. Smith's land.

Notice that the 1 part that is to go to Mr. Henley's heir is $\frac{1}{4}$ of Mr. Henley's land, but it is also $\frac{1}{3}$ of Mr. Smith's land. So there are two different *wholes* in this exercise: Mr. Henley's land, that the $\frac{3}{4}$ and the $\frac{1}{4}$ in the exercise are associated with; and Mr. Smith's land, that the answer $\frac{1}{3}$ is associated with.

5. Because 1 serving of juice provides $\frac{3}{2}$ of the daily value of vitamin C, therefore 1 serving of juice represents 3 parts, when the daily value of vitamin C is divided into 2 equal parts. The daily value of vitamin C is represented in 2 of those parts, as shown in Figure 2.36. If you drink

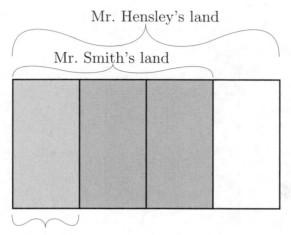

Mr. Hensley's land

Mr. Smith's land

1/4 of Mr. Hensley's land is 1/3 of
Mr. Smith's land

Figure 2.35: $\frac{1}{4}$ of Mr. Hensley's Land is $\frac{1}{3}$ of Mr. Smith's Land

$\frac{2}{3}$ of a serving of juice, you will get those 2 parts of the daily value of vitamin C, so you will get the full daily value of vitamin C in $\frac{2}{3}$ of a serving of juice.

There are two different *wholes* associated with the fractions in this problem: a serving of juice and the daily value of vitamin C. The whole associated with the $\frac{3}{2}$ in the exercise is the "daily value of vitamin C". The whole associated with the $\frac{2}{3}$ in the exercise is "a serving of juice".

6. See Figure 2.37.

Problems for Section 2.4 on the Meaning of Fractions

1. Michael says that the dark marbles in Figure 2.38 can't represent $\frac{1}{3}$ because they are 5 marbles. Write a short paragraph discussing the source of Michael's confusion and what you might tell Michael to help him.

2. Kaitlyn gave $\frac{1}{2}$ of her candy bar to Arianna. Arianna gave $\frac{1}{3}$ of the candy she got from Kaitlyn to Cameron. What fraction of a candy bar did Cameron get? Draw pictures and use your pictures to help you

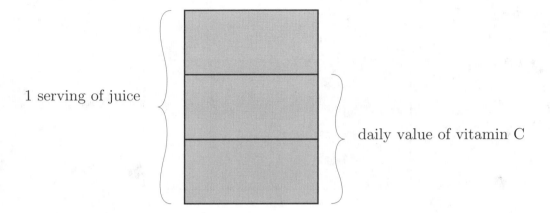

1 serving of juice

daily value of vitamin C

Figure 2.36: 1 Serving of Juice Provides $\frac{3}{2}$ of the Daily Value of Vitamin C

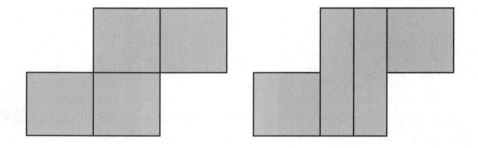

Figure 2.37: Two Ways to Divide Into 4 Equal Parts

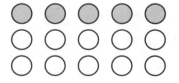

Figure 2.38: Some Marbles

solve this problem. Explain clearly how your pictures help you solve the problem. For each fraction in this problem, and in your solution, describe the *whole* that this fraction is associated with. Then write a paragraph discussing the roles of these different *wholes* in the solution to the problem.

3. Susan was supposed to use $\frac{5}{4}$ of a cup of butter in her recipe but she only used $\frac{3}{4}$ of a cup of butter. What fraction of the butter that she should have used did Susan actually use? Draw pictures and use your pictures to help you solve this problem. Explain clearly how your pictures help you solve the problem. For each fraction in this problem, and in your solution, describe the *whole* that this fraction is associated with.

4. A container is filled with $\frac{5}{2}$ of a cup of cottage cheese. What fraction of the cottage cheese in the container should you eat if you want to eat 1 cup of cottage cheese? Draw pictures and use your pictures to help you solve this problem. Explain clearly how your pictures help you solve the problem. For each fraction in this problem, and in your solution, describe the *whole* that this fraction is associated with.

5. Make up a story problem or situation where *one* object (or collection, or quantity) is *both* $\frac{1}{2}$ of something and $\frac{1}{3}$ of something else.

6. Write a paragraph discussing which of the items on the list below are good for showing improper fractions, and which are not as good for showing improper fractions. Explain your choices.

   string
   a cake
   a box of cereal and some cup measures
   apples
   something else—your choice

7. NanHe used hexagons, rhombuses, and triangles like the ones in Figure 2.39 to make a design. NanHe counted how many of each shape she used in her design and determined that $\frac{4}{11}$ of the shapes she used were hexagons, $\frac{5}{11}$ were rhombuses, and $\frac{2}{11}$ were triangles. You may combine your answers to parts (a) and (b).

   (a) Even though $\frac{4}{11}$ of the shapes NanHe used were hexagons, does this mean that the hexagons in NanHe's design take up $\frac{4}{11}$ *of the*

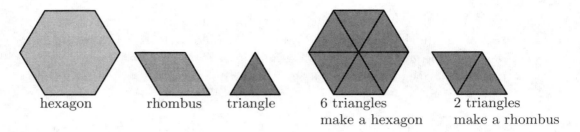

Figure 2.39: Pattern Blocks

> *area* of her design? If not, what fraction of the area of NanHe's design do the hexagons take up?

(b) Write a short paragraph on how the notion of *equal parts* arises in this problem.

8. A county has two elementary schools, both of which have an after school program for the convenience of working parents. In one school, $\frac{1}{3}$ of the children attend the after school program; in the other school, $\frac{2}{3}$ of the children attend the after school program. There are a total of 1500 children in the two schools combined.

(a) If the first school has 600 students and the second school has 900 students, then what is the fraction of elementary school students in the county who attend the after school program? What if it's the other way around and the first school has 900 students and the second has 600 students?

(b) Using the data given at the beginning of the problem (before part (a)), make up two more examples of numbers of students in each school (these don't have to be entirely realistic numbers). For each of your examples, determine the fraction of elementary school children in the county who attend the after school program.

(c) Jamie says that if $\frac{1}{3}$ of the children in the first school attend the after school program, and if $\frac{2}{3}$ of the children in the second school attend the after school program, then $\frac{1}{2}$ of the children from the two schools combined attend the after school program. Jamie explains that this is because $\frac{1}{2}$ is half-way between $\frac{1}{3}$ and $\frac{2}{3}$. Is

Jamie's answer of $\frac{1}{2}$ *always* correct? Is Jamie's answer of $\frac{1}{2}$ *ever* correct, if so, under what circumstances?

2.5 Equivalent Fractions

When we write a whole number as a decimal, there is only one way to do so without the use of a decimal point. The only way to write 1234 as a decimal without a decimal point is 1234. But in the case of fractions, *every fraction is equal to infinitely many other fractions.* For example,

$$\frac{1}{2} = \frac{2}{4} = \frac{3}{6} = \frac{4}{8} = \frac{5}{10} = \cdots$$

In general,

$$\frac{A}{B} = \frac{A \cdot 2}{B \cdot 2} = \frac{A \cdot 3}{B \cdot 3} = \frac{A \cdot 4}{B \cdot 4} = \frac{A \cdot 5}{B \cdot 5} = \cdots.$$

In this section we will see why every fraction is equal to infinitely many other fractions, and we will study some consequences of this fact.

Class Activity 2H: Equivalent Fractions

Why Every Fraction is Equal to Infinitely Many Other Fractions

Let's explain why

$$\frac{3}{4}$$

of a pie the same amount of pie as

$$\frac{3 \cdot 5}{4 \cdot 5} = \frac{15}{20}$$

of the same pie. We can use the meaning of fractions of objects to explain this. First, $\frac{3}{4}$ of the pie is the amount formed by 3 parts when the pie is divided into 4 equal parts. Divide each of those 4 equal parts into 5 small, equal parts, as shown in Figure 2.40. Now the pie consists of a total of $4 \cdot 5 = 20$ small parts. The 3 parts representing $\frac{3}{4}$ of the pie have each been subdivided into 5 equal parts, therefore these 3 parts have become $3 \cdot 5 = 15$ smaller parts, as shown in the shaded portions of Figure 2.40. So 3 of the

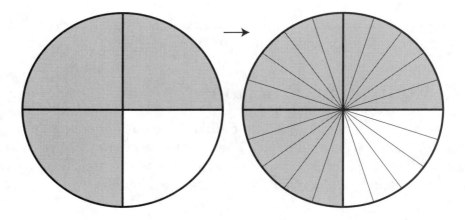

Figure 2.40: Subdivide Each Part Into 5 Parts

original 4 parts of pie is the same amount of pie as $3 \cdot 5$ smaller parts of the total $4 \cdot 5$ smaller parts. This tells us that

$$\frac{3}{4}$$

of a pie is the same amount of pie as

$$\frac{3 \cdot 5}{4 \cdot 5} = \frac{15}{20}$$

of the pie.

There wasn't anything special about the numbers 3, 4, and 5 used here—the same reasoning will apply with other counting numbers substitued for these numbers. Therefore in general, if N is any counting number,

$$\frac{A}{B}$$

of a pie (or any other object or collection of objects) is the same amount of pie (or object or collection) as

$$\frac{A \cdot N}{B \cdot N}$$

of the pie (or object or collection), in other words,

$$\frac{A}{B} = \frac{A \cdot N}{B \cdot N}.$$

Another nice way to explain why

$$\frac{A}{B} = \frac{A \cdot N}{B \cdot N}$$

is to think of this as *multiplying by 1*:

$$\begin{aligned} \frac{A}{B} &= \frac{A}{B} \cdot 1 \\ &= \frac{A}{B} \cdot \frac{N}{N} \\ &= \frac{A \cdot N}{B \cdot N}. \end{aligned}$$

This explanation for why $\frac{A}{B} = \frac{A \cdot N}{B \cdot N}$ relies on the conceptually more complex concept of fraction multiplication, whereas the previous explanation relies only on the meaning of fractions.

Class Activity 2I: When Can We "Cancel" to Get an Equivalent Fraction?

Flexibility in Writing Fractions

Because every fraction is equal to infinitely many other fractions, therefore you have a lot of flexibility when you work with fractions. Starting with a fraction, say $\frac{3}{4}$, it is always equal to many other fractions with larger denominators:

$$\frac{3}{4} = \frac{3 \cdot 7}{4 \cdot 7} = \frac{21}{28},$$

$$\frac{3}{4} = \frac{3 \cdot 25}{4 \cdot 25} = \frac{75}{100},$$

and many others.

In some cases, a fraction is equal to another fraction with a smaller denominator:

$$\frac{8}{12} = \frac{2 \cdot 4}{3 \cdot 4} = \frac{2}{3},$$

$$\frac{170}{190} = \frac{17 \cdot 10}{19 \cdot 10} = \frac{17}{19}.$$

When working with two fractions simultaneously it is often desirable to give them **common denominators**, which just means the *same* denomi- **common denominators**

nators. The fractions $\frac{3}{5}$ and $\frac{2}{7}$ can be written with the common denominator $5 \cdot 7 = 35$:

$$\frac{3}{5} = \frac{3 \cdot 7}{5 \cdot 7} = \frac{21}{35}$$
$$\frac{2}{7} = \frac{2 \cdot 5}{7 \cdot 5} = \frac{10}{35}.$$

The fractions $\frac{3}{5}$ and $\frac{2}{7}$ have many other common denominators as well, such as 70 and 105, but 35 is the smallest one.

For any two fractions, multiplying the denominators always produces a common denominator, however, it may not be the smallest one. The number 24 is a common denominator for the fractions $\frac{3}{8}$ and $\frac{5}{6}$, whereas $8 \cdot 6 = 48$.

Class Activity 2J: Common Denominators

Class Activity 2K: Using Fractions Flexibly by Writing Them in Different Ways

The Simplest Form of a Fraction

Every fraction is equal to infinitely many other fractions, but in a collection of fractions that are equal to each other, there is one that is the simplest.

A fraction of whole numbers $\frac{A}{B}$ (where B is not zero) is said to be in **simplest form** (or in **lowest terms**) if there is no whole number other than 1 that divides both A and B evenly.

The fraction

$$\frac{3}{4}$$

is in simplest form because no whole number other than 1 divides both 3 and 4 evenly. The fraction

$$\frac{30}{35}$$

is not in simplest form because the number 5 divides both 30 and 35 evenly. Notice, however, that we can put the fraction $\frac{30}{35}$ in simplest form as follows:

$$\frac{30}{35} = \frac{6 \cdot 5}{7 \cdot 5} = \frac{6}{7}.$$

In general, if

$$\frac{A}{B}$$

is a fraction of whole numbers that is not in simplest form, then you can put it in simplest form as follows. First find the largest whole number N that divides both A and B evenly. Write A and B in the form $C \cdot N$ and $D \cdot N$ respectively, where C and D are whole numbers. Then

$$\frac{A}{B} = \frac{C \cdot N}{D \cdot N} = \frac{C}{D},$$

and $\frac{C}{D}$ is the simplest form of $\frac{A}{B}$.

To write $\frac{18}{24}$ in simplest form, notice that 6 is the largest whole number that divides both 18 and 24 evenly.

$$\frac{18}{24} = \frac{3 \cdot 6}{4 \cdot 6} = \frac{3}{4},$$

therefore $\frac{3}{4}$ is the simplest form of $\frac{18}{24}$.

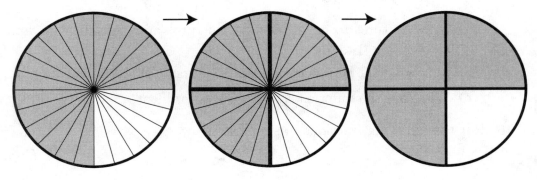

Figure 2.41: Putting $\frac{18}{24}$ in Simplest Form

If you think in terms of pies, putting a fraction in simplest form is like making larger pie pieces out of many smaller ones, as shown in Figure 2.41. In other words, putting a fraction in simplest form reverses the procedure of subdividing pie pieces that was demonstrated earlier in Figure 2.40.

Exercises for Section 2.5 on Equivalent Fractions

1. Use the meaning of fractions to explain why

$$\frac{4}{5} = \frac{4 \cdot 3}{5 \cdot 3} = \frac{12}{15}.$$

2. Write $\frac{3}{8}$ and $\frac{5}{6}$ with common denominators in three different ways.

3. Kelsey inherited $\frac{3}{5}$ of her parents' land. Now Kelsey wants to give her daughter some of this land. What fraction of her land should Kelsey give her daughter so that Kelsey will be left with $\frac{1}{2}$ of her parents' original land?

   (a) Draw pictures to help you solve this problem. Explain why your answer is correct.

   (b) In solving this problem, how do $\frac{3}{5}$ and $\frac{1}{2}$ appear in different forms?

   (c) What are the different *wholes* in this problem?

4. Ken ordered $\frac{3}{4}$ of a ton of gravel. Ken wants $\frac{1}{4}$ of his order of gravel delivered now, and $\frac{3}{4}$ delivered later. What fraction of a ton of gravel should Ken get delivered now?

   (a) Draw pictures to help you solve this problem. Explain why your answer is correct.

   (b) In solving the problem, how does $\frac{3}{4}$ appear in a different form?

   (c) What are the different *wholes* in this problem?

5. Put the following fractions in simplest form:

$$\frac{45}{72}, \quad \frac{24}{36}, \quad \frac{56}{88}.$$

Answers to Exercises for Section 2.5 on Equivalent Fractions

1. If you have a pie that is divided into 5 equal parts, and 4 are shown shaded, then if you divide each of the 5 parts into 3 smaller parts, the shaded amount will then consist of $4 \cdot 3$ small parts, and the whole pie will consist of $5 \cdot 3$ small parts. Thus the shaded part of the pie can be described both as $\frac{4}{5}$ of the pie and as

$$\frac{4 \cdot 3}{5 \cdot 3} = \frac{12}{15}$$

of the pie.

2. There are infinitely many ways to write $\frac{3}{8}$ and $\frac{5}{6}$ with common denominators. Here are three:

$$\frac{3}{8} = \frac{3 \cdot 6}{8 \cdot 6} = \frac{18}{48},$$
$$\frac{5}{6} = \frac{5 \cdot 8}{6 \cdot 8} = \frac{40}{48}.$$

$$\frac{3}{8} = \frac{3 \cdot 3}{8 \cdot 3} = \frac{9}{24},$$
$$\frac{5}{6} = \frac{5 \cdot 4}{6 \cdot 4} = \frac{20}{24}.$$

$$\frac{3}{8} = \frac{3 \cdot 9}{8 \cdot 9} = \frac{27}{72},$$
$$\frac{5}{6} = \frac{5 \cdot 12}{6 \cdot 12} = \frac{60}{72}.$$

3. (a) See Figure 2.42. If Kelsey's parents' land is divided into 10 equal parts, then Kelsey's land is $\frac{6}{10}$ of her parents' land and the land she wants to keep is $\frac{5}{10}$ of her parents' land. The land Kelsey wants to give to her daughter is 1 part out of the 6 equal parts making up Kelsey's current land. Therefore Kelsey's daughter will get $\frac{1}{6}$ of Kelsey's land.

   (b) In solving this problem, $\frac{3}{5}$ becomes

   $$\frac{3 \cdot 2}{5 \cdot 2} = \frac{6}{10}$$

   and $\frac{1}{2}$ becomes

   $$\frac{1 \cdot 5}{2 \cdot 5} = \frac{5}{10}.$$

   (c) Kelsey's parents' land is one of the *wholes* in this problem— Kelsey's current land is $\frac{3}{5}$ of this whole, and so is the $\frac{1}{2}$ that Kelsey wants to keep. Kelsey's current land is also treated as a *whole* in this problem because we describe the land that Kelsey's daughter will get as a fraction of this whole.

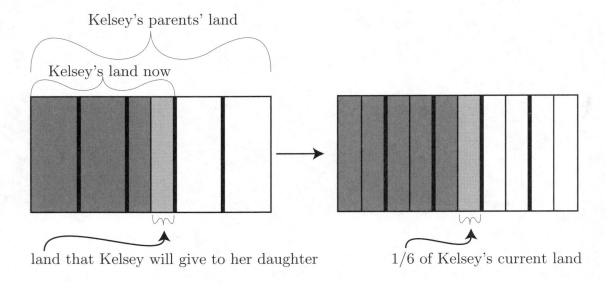

Figure 2.42: What Fraction of Kelsey's Land Will Her Daughter Get?

4. Ken ordered $\frac{3}{4}$ of a ton of gravel. Ken wants $\frac{1}{4}$ of his order of gravel delivered now, and $\frac{3}{4}$ of his order delivered later. What fraction of a ton of gravel should Ken get delivered now?

   (a) In Figure 2.43, 1 ton of gravel is represented by a circle, and Ken's order is represented by $\frac{3}{4}$ of a circle. On the right, $\frac{1}{4}$ of Ken's order is represented by dividing each of the 3 parts in the $\frac{3}{4}$ into 4 equal small pieces, and picking one small piece from each of the 3 parts. These small pieces are shown shaded darkly. There are 16 small pieces in the full circle, and the 3 shaded darkly represent $\frac{1}{4}$ of Ken's order. Therefore $\frac{1}{4}$ of Ken's order is $\frac{3}{16}$ of a ton.

   (b) In solving this problem, the $\frac{3}{4}$ of a ton of gravel also appears as

   $$\frac{3 \cdot 4}{4 \cdot 4} = \frac{12}{16}$$

   of a ton of gravel.

   (c) A full ton of gravel is one of the *wholes* in this problem. Ken's order of gravel ($\frac{3}{4}$ of a ton) is another *whole*, because in this problem we determine what $\frac{1}{4}$ of this whole is as a fraction of 1 ton of gravel.

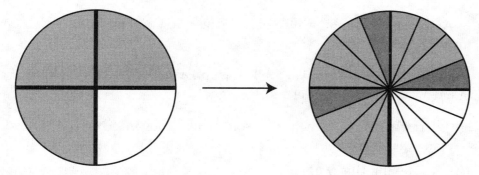

whole circle: 1 ton of gravel

shaded part: Ken's order of 3/4 ton

1/4 of the 3 large shaded pieces is shaded darkly -- it is 3/16 of a ton

Figure 2.43: What Fraction of a Ton Should Ken Get Now?

5.

$$\frac{45}{72} = \frac{5 \cdot 9}{8 \cdot 9} = \frac{5}{9},$$
$$\frac{24}{36} = \frac{2 \cdot 12}{3 \cdot 12} = \frac{2}{3},$$
$$\frac{56}{88} = \frac{7 \cdot 8}{11 \cdot 8} = \frac{7}{11}.$$

Problems for Section 2.5 on Equivalent Fractions

1. Use the meaning of fractions to explain why

$$\frac{6}{5} = \frac{6 \cdot 2}{5 \cdot 2}.$$

2. Anna says:

$$\frac{2}{3} = \frac{6}{7}$$

because starting with $\frac{2}{3}$, you get $\frac{6}{7}$ by adding 4 to the top and the bottom. If you do the same thing to the top and the bottom, the fractions must be equal.

Is Anna right? If not, why not? Discuss how might Anna have come up with her rule.

3. Ken got $\frac{2}{3}$ of a ton of gravel even though he ordered $\frac{3}{4}$ of a ton of gravel. What fraction of his order did Ken get?

   (a) Draw pictures to help you solve this problem. Explain why your answer is correct.

   (b) In solving this problem, how do $\frac{2}{3}$ and $\frac{3}{4}$ appear in different forms?

   (c) What are the different *wholes* in this problem?

4. One serving of DietMuck is $\frac{2}{3}$ of a cup. Jane wants to eat $\frac{2}{3}$ of a serving of DietMuck. What fraction of a cup of DietMuck should Jane eat?

   (a) Draw pictures to help you solve this problem. Explain why your answer is correct.

   (b) In solving the problem, how does $\frac{2}{3}$ appear in a different form?

   (c) What are the different *wholes* in this problem?

5. Ted put $\frac{3}{4}$ of a cup of chicken stock in his soup. Later, Ted added another $\frac{2}{3}$ of a cup of chicken stock to his soup. All together, what fraction of a cup of chicken stock did Ted put into his soup?

   (a) Draw pictures to help you solve this problem. Explain why your answer is correct.

   (b) In solving the problem, how do $\frac{3}{4}$ and $\frac{2}{3}$ appear in different forms?

   (c) What are the different *wholes* in this problem?

6. Draw a picture showing $\frac{12}{16}$ of a pie. Use your picture of $\frac{12}{16}$ of a pie to help you explain why
$$\frac{12}{16} = \frac{3}{4}.$$

   Starting with $\frac{12}{16}$ of a pie, what do you need to do to get $\frac{3}{4}$ of a pie?

7. Becky moves into an apartment with two friends on February 1. Becky's friends have been in the apartment since January 1. The electric bill comes every two months and the next one will be for the electricity used in January and February. The bill is *not* broken down by month.

What fraction of the January/February electric bill should Becky pay and what fraction should her two friends pay if they want to divide the bill fairly? Explain your reasoning clearly.

2.6 Fractions as Numbers

So far we have only discussed fractions of objects, collections or quantities, but we haven't yet looked at fractions abstractly, as numbers in their own right.

We create the notion of the whole numbers by abstracting from our experiences with objects:

2 apples, 7 balls, 25 people, ...

is really like saying (somewhat awkwardly):

2 *of* apple, 7 *of* ball, 25 *of* person, ...,

which abstracts to the notion of number:

2, 7, 25, ...

In the same way, we create the notion of fractions as numbers by abstracting from fractions of objects:

$\frac{2}{3}$ *of* a pie, $\frac{11}{10}$ *of* an acre of land, $\frac{7}{8}$ *of* the population of the United States, ...

abstract to

$\frac{2}{3}, \frac{11}{10}, \frac{7}{8}, \ldots$

But even when fractions are viewed abstractly as numbers, they are still *of a whole*: just as 5 is "five ones," so too $\frac{3}{4}$ is "$\frac{3}{4}$ of 1."

By viewing fractions as locations on number lines we will be able to see even more clearly how fractions are numbers in their own right.

Fractions on Number Lines

Class Activity 2L: Fractions Of Line Segments

We will now see how fractions can be plotted on a number line in a natural way, using the ideas of the previous class activity.

Any fraction $\frac{A}{B}$, where A and B are whole numbers and B is not zero can be located on a number line according to the rule for locating decimals on number lines given in Section 2.3. Thus $\frac{A}{B}$ is located to the right of zero at a distance of $\frac{A}{B}$ units from 0. This means that $\frac{A}{B}$ should be at the right end of a line segment whose left end is at 0 and whose length is $\frac{A}{B}$ *of one unit.*

Where is $\frac{5}{3}$ on a number line? $\frac{5}{3}$ should be at the right end of a line sement whose left end is at 0 and whose length is $\frac{5}{3}$ of a unit. To construct such a line segment, use the meaning of $\frac{5}{3}$: divide the line segment from 0 to 1 into 3 equal parts, and make a new line segment consisting of 5 copies of those parts, as shown in Figure 2.44.

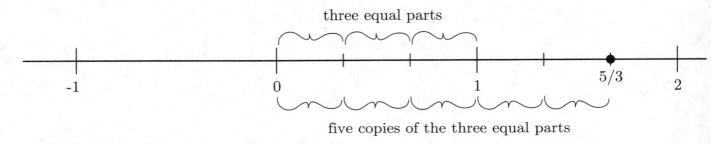

Figure 2.44: Plotting $\frac{5}{3}$ on a Number Line

Decimal Representations Of Fractions

Because fractions can be plotted on number lines, and because every location on a number line corresponds to a decimal, therefore it must be possible to represent fractions as decimals. In fact, it is easy to write a fraction $\frac{A}{B}$ as a decimal: just divide A by B.

$$\frac{A}{B} = A \div B.$$

We will see why it makes sense to write a fraction as a decimal by dividing in Chapter 5.

So,

$$\frac{5}{16} = 5 \div 16 = .3125,$$
$$\frac{1}{12} = 1 \div 12 = .083333333\ldots,$$
$$\frac{2}{7} = 2 \div 7 = .285714285714\ldots.$$

Since a fraction can be written as decimal by dividing the numerator by the denominator, it has become common practice to write a fraction when one actually means division, in other words, it is common to write

$$\frac{A}{B} \quad \text{or} \quad A/B$$

to mean

$$A \div B.$$

In fact, some calculators use the fraction symbol / instead of ÷.

Exercises for Section 2.6 on Fractions as Numbers

1. The line segment below has length $\frac{3}{7}$ unit. Use a marked rubber band to construct a line segment of length $\frac{1}{7}$ unit. Do this without first constructing a segment of length 1 unit. Explain how you know your segment is the correct length.

3/7 units

2. The line segment below has length $\frac{9}{8}$ unit. Use a marked rubber band to construct a line segment of length $\frac{3}{4}$ unit. Do this without first constructing a segment of length 1 unit. Explain how you know your segment is the correct length.

9/8 units

3. The line segment below has length $\frac{3}{5}$ unit. Use a marked rubber band to construct a line segment of length $\frac{2}{3}$ unit. Do this without first constructing a segment of length 1 unit. Explain how you know your segment is the correct length.

3/5 units

4. Use a marked rubber band to help you plot $\frac{11}{8}$ on the number line in Figure 2.45.

Figure 2.45: A Number Line

5. Use a marked rubber band to help you plot $\frac{29}{3}$ on the number line in Figure 2.46.

Figure 2.46: A Number Line

Answers to Exercises for Section 2.6 on Fractions as Numbers

1. Because the line segment is $\frac{3}{7}$ of a unit long, it must consist of 3 parts, each of which is $\frac{1}{7}$ of a unit long. Therefore if the line segment is divided into 3 equal parts, one of those parts will be $\frac{1}{7}$ of a unit long, as shown in Figure 2.47.

2. Because the original line segment is $\frac{9}{8}$ units long, it consists of 9 equal parts, each of which is $\frac{1}{8}$ of a unit long. Since

$$\frac{3}{4} = \frac{3 \cdot 2}{4 \cdot 2} = \frac{6}{8},$$

we should create a line segment consisting of 6 of the $\frac{1}{8}$ unit long parts, as shown in Figure 2.48.

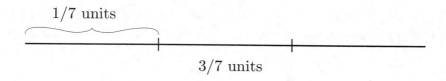

Figure 2.47: From $\frac{3}{7}$ Units to $\frac{1}{7}$ Units

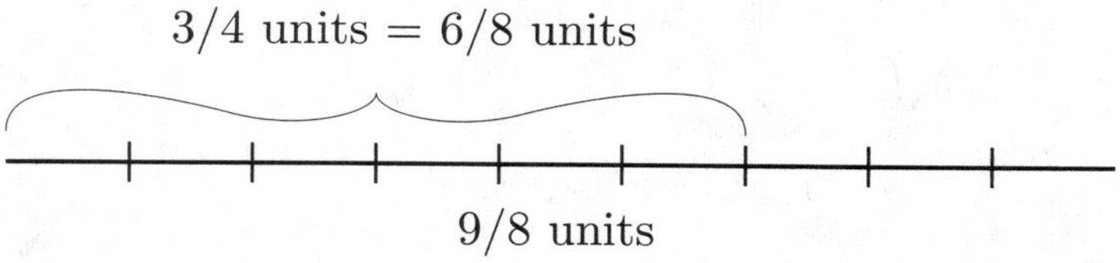

3/4 units = 6/8 units

9/8 units

Figure 2.48: From $\frac{9}{8}$ Units to $\frac{3}{4}$ Units

3. Because

$$\frac{3}{5} = \frac{3 \cdot 3}{5 \cdot 3} = \frac{9}{15}$$

and

$$\frac{2}{3} = \frac{2 \cdot 5}{3 \cdot 5} = \frac{10}{15},$$

therefore the original line segment of length $\frac{3}{5}$ consists of 9 equal parts, each of which is $\frac{1}{15}$ of a unit long, and the desired line segment of length $\frac{2}{3}$ units will consist 10 copies of those parts, as shown in Figure 2.49.

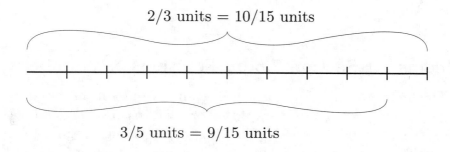

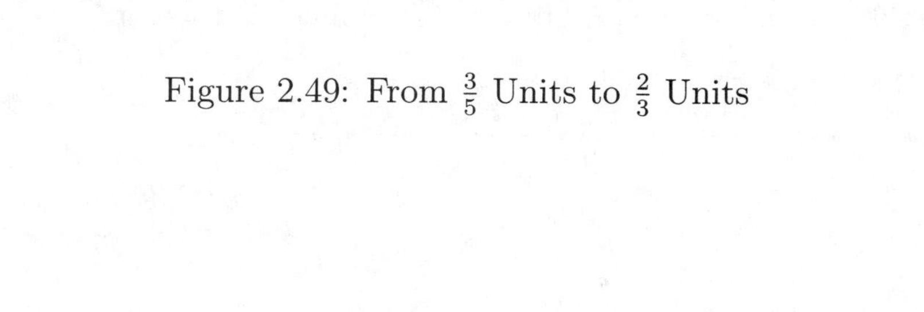

Figure 2.49: From $\frac{3}{5}$ Units to $\frac{2}{3}$ Units

4. To plot $\frac{11}{8}$ on a number line, divide the 1-unit segment from 0 to 1 into 8 equal parts, and measure off 11 of those parts, as shown in Figure 2.50.

11 parts when 1 unit is divided into 8 equal parts

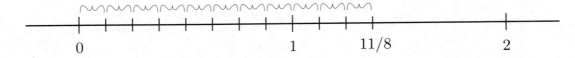

Figure 2.50: Plotting $\frac{11}{8}$

5. The fraction $\frac{29}{3}$ is located 29 parts away from 0, where each part is $\frac{1}{3}$ of a unit long, i.e., 3 parts are 1 unit long. 9 sets of 3 parts will use up 27 parts, and these 27 parts will end at 9 on the number line. 2 more parts will get to the location of $\frac{29}{3}$ on the number line, as shown in Figure 2.51.

9 sets of 3 make 27 parts   2 more parts make 29

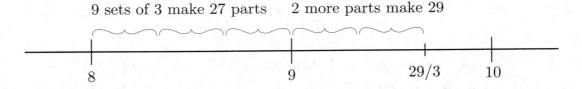

Figure 2.51: Plotting $\frac{29}{3}$

Problems for Section 2.6 on Fractions as Numbers

1. The line segment below has length $\frac{2}{7}$ unit. Copy this line segment onto a piece of paper. Use a marked rubber band to construct a line segment of length $\frac{1}{3}$ unit. Do this without first constructing a segment of length 1 unit. Explain how you know your segment is the correct length.

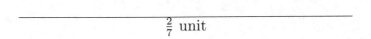

$\frac{2}{7}$ unit

2. The line segment below has length $\frac{4}{5}$ unit. Copy this line segment onto a piece of paper. Use a marked rubber band to construct a line segment of length $\frac{1}{2}$ unit. Do this without first constructing a segment of length 1 unit. Explain how you know your segment is the correct length.

$\frac{4}{5}$ unit

3. The line segment below has length $\frac{3}{5}$ unit. Copy this line segment onto a piece of paper. Use a marked rubber band to construct a line segment of length $\frac{1}{2}$ unit. Do this without first constructing a segment of length 1 unit. Explain how you know your segment is the correct length.

$\frac{4}{5}$ unit

4. Copy the number line in Figure 2.52 onto a piece of paper. Use a marked rubber band to help you plot $\frac{13}{3}$ on this number line.

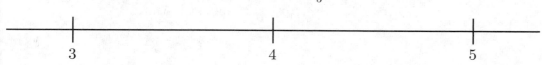

Figure 2.52: Plot $\frac{13}{3}$

5. Copy the number line in Figure 2.53 onto a piece of paper. Use a marked rubber band to help you plot $\frac{47}{6}$ on this number line.

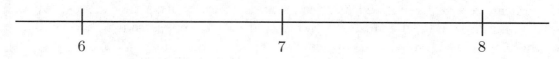

Figure 2.53: Plot $\frac{47}{6}$

6. Erin says to label the tick mark indicated in Figure 2.54 as 2.2. Is Erin right or not? If not, why not, and how can the tick mark be labeled properly?

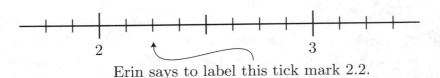

Erin says to label this tick mark 2.2.

Figure 2.54: How To Label The Tick Mark?

7. Liam says to label the tick mark indicated in Figure 2.55 as 1.7. Is Liam right or not? If not, why not, and how can the tick mark be labeled properly?

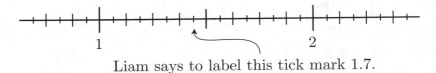

Liam says to label this tick mark 1.7.

Figure 2.55: How to Label the Tick Mark?

2.7   Percent

Almost any time we open a newspaper or walk in a store, we encounter percentages. A percentage is a number that expresses a relationship between two other numbers. In this way, percentages are like fractions: just as a fraction is *of* an associated whole, a percentage is *of* some number. In fact, percentages can be viewed as special kinds of fractions, namely those with denominator 100. Because every fraction can be expressed as a decimal, therefore every percentage can also be expressed as a decimal.

The Meaning of Percent

The word **percent**, which is usually indicated with the symbol %, means *of each hundred* (per = of each, or for each, cent = hundred). For example, 76% means *76 of each hundred.* Therefore, to find 76% of 1280, we should see how many hundreds are in 1280, and multiply that number by 76. The number of hundreds in 1280 is

$$1280 \div 100,$$

therefore 76% of 1280 is

$$76 \times (1280 \div 100),$$

which can also be expressed as

$$\frac{76}{100} \cdot 1280,$$

or even

$$.76 \cdot 1280.$$

So 76% of 1280 is 972.8.

Another example:

$$
\begin{aligned}
3\% \text{ of } 234 \ &= \ \frac{3}{100} \cdot 234 \\
&= \ .03 \cdot 234 \\
&= \ 7.02.
\end{aligned}
$$

In general, $P\%$ of a quantity Q is

$$\frac{P}{100} \cdot Q.$$

Because $P\%$ of a number is $\frac{P}{100}$ times that number, therefore it makes sense to say that

$$P\% = \frac{P}{100},$$

so that

$$76\% = \frac{76}{100} = .76,$$

and

$$125\% = \frac{125}{100} = 1.25.$$

We can therefore interpret the word *percent* to mean *hundredths*, and we can view percents as decimals, or as special kinds of fractions, namely ones that have denominator 100. When we consider percents as fractions, we can use the meaning of fractions, so that 85% of some object means the amount formed by 85 parts when the object is divided into 100 equal parts.

Percents, Fractions, and Pictures

To get a feel for percentages it is useful to know how to express some common percentages as fractions in simplest form:

$$25\% = \frac{25}{100} = \frac{1}{4}, \quad 50\% = \frac{50}{100} = \frac{1}{2}, \quad 75\% = \frac{75}{100} = \frac{3}{4},$$

$$10\% = \frac{10}{100} = \frac{1}{10}, \quad 20\% = \frac{20}{100} = \frac{1}{5}, \quad 5\% = \frac{5}{100} = \frac{1}{20}.$$

See Figure 2.56.

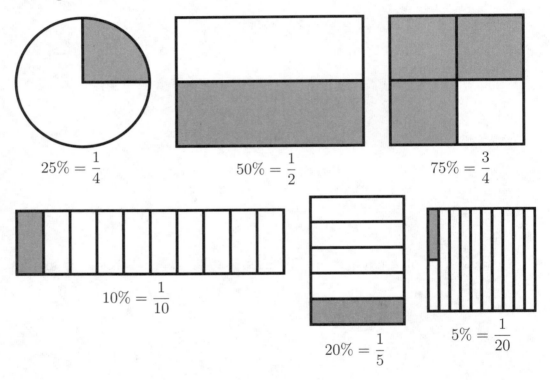

Figure 2.56: Pictures of Percentages

It is also useful to be able to think about percentages in terms of nearby percentages and fractions. For example:

55% is a little more than $\frac{1}{2}$, and is half way between 50% and 60%, or

15% is 10% less than $\frac{1}{4}$.

Some of the class activities, exercises, and problems ask you to use pictures to solve percent problems. Although drawing pictures is usually not an efficient way of solving percent problems, pictures can help you develop a better understanding of percents. Just as it would not be reasonable to have to pull out bundled toothpicks whenever we work with decimals, it would also not be reasonable to expect to do all percent calculations with pictures. However, both physical objects and pictures can be valuable learning aids.

Class Activity 2M: Percent

Three Types of Percent Problems

A basic relationship involving percents can be put in the form

P percent of Q is R.

In equation form this is:

$$P\% \cdot Q = R$$

or

$$\frac{P}{100} \cdot Q = R.$$

Therefore there are three basic kinds of percent problems: one for each of the cases where one of the amounts P, Q, and R is unknown, and to be determined, and the other two amounts are known.

Finding the Result R when the Percent P and the Initial Quantity Q are Known

The most straightforward kind of problem involving percentages is one in which the percent P and the quantity Q are given, and the resulting amount R must be calculated:

Susie must pay 6% tax on her purchase of $43.95. How much tax must Susie pay?

Susie must pay

$$.06 \cdot \$43.95 = \$2.64$$

in tax.

Finding the Percent P when the Initial Quantity Q and the Result R are Known

In some problems, the initial amount Q and the resulting amount R are known, but the percent P must be calculated:

> John earned $\$47,600$ last year and gave $\$375$ to charity. What percent of his income did John give to charity?

In this problem, we know that $P\%$ of $47,600$ is 375 and we want to find the percent P. So

$$P\% \cdot 47,600 \;=\; 375,$$

therefore

$$P\% = \frac{375}{47,600} = .00788 = .788\%$$

which means that John gave about $.8\%$ (8 tenths of one percent) of his income to charity.

If Nellie leaves a $\$9$ tip on a meal that cost $\$57.45$, what percent was Nellie's tip? If $P\%$ is this percentage, then

$$P\% \cdot 57.45 = 9,$$

so

$$P\% = \frac{9}{57.45} = 9 \div 57.45 = .157 = 15.7\%.$$

Therefore Nellie left a tip of nearly 16%.

Notice that to solve this type of problem, you really just divide the resulting amount R by the initial amount Q (and then multiply by 100 to write the answer as a percentage).

Finding the original quantity Q when the percent P and the resulting amount R are known

In another basic type of percent problem, we know the percent P and the resulting amount R, and we must calculate the initial quantity Q:

> Nancy paid $\$1164$ in property taxes on her home. The taxes Nancy paid represent $.92\%$ of the value of her home. What is the value of Nancy's home?

Let's let N stand for the value of Nancy's home. Then

$$\frac{.92}{100} \cdot N = 1164$$
$$\text{so}$$
$$.0092 \cdot N = 1164$$
$$N = 1164 \div .0092$$
$$N = 126,522,$$

which means Nancy's home is worth $126,522.

Class Activity 2N: Calculations with Percents

2.7.1 Exercises for Section 2.7 on Percent

1. For each of the following shapes in Figure 2.57, determine what percent of the shape is shaded. Give your answer rounded to the nearest multiple of 5 (i.e., 5, 10, 15, 20, ...).

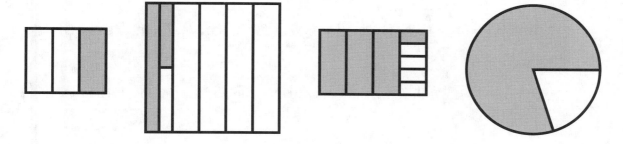

Figure 2.57: What Percent Is Shaded In Each?

2. Ted left a tip of $7.00 on a meal. If this represents 20% of the cost of the meal (Ted is a good tipper), then how much did the meal cost? Use a line segment or a number line to solve this problem. Also solve the problem numerically.

3. (a) 1067 is 136% of what number?

   (b) What percent of 843 is 1255?

4. At one point, 15% of the US budget went to pay interest on the federal debt, and the interest on the federal debt was $241 billion. What was the US budget at that time?

5. If $12.3 million is 75% of the budget, then how much is the full budget? First solve the problem by drawing a picture. Explain how your picture helps you solve the problem. Then solve the problem numerically.

6. There were 4800 gallons of water in a tank. Some of the water was drained out, leaving 65% of the original amount of water in the tank. How many gallons of water are in the tank? First solve the problem by drawing a picture. Explain how your picture helps you solve the problem. Then solve the problem numerically.

7. George was given 9 grams of medicine, but he was supposed to have received 20 grams. What percent of the amount of medicine that he should have received did George actually receive? First solve the problem by drawing a picture. Explain how your picture helps you solve the problem. Then solve the problem numerically.

2.7.2   Answers to Exercises for Section 2.7 on Percent

1. See Figure 2.58.

2. $20\% = \frac{1}{5}$. If $7.00 represents one fifth of the cost of the meal, then the meal must cost 5 times $7.00, which is $35.00. Another way to solve this is:

$$
\begin{array}{rcl}
20\% \text{ of } ? & = & 7 \\
.20 \times ? & = & 7 \\
? & = & 7 \div .20 \\
? & = & 35
\end{array}
$$

See Figure 2 for a pictorial solution to the problem.

3. (a) Stating the question numerically:

$$\left(\frac{136}{100}\right) \times ? = 1067,$$

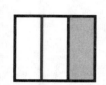

1/3 is about 33%, which rounds to 35%

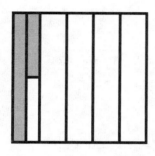

Each big strip is 20%, so the shaded part is 15%

Each big strip is 25%. The little strips are 1/5 of that, so 5%. So 80% is shaded.

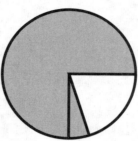

The large portion is 75%. The small piece must be about 5% because 10% would be close to half of the remaining 25% of the pie. So about 80% is shaded.

Figure 2.58: Percent Shaded

| $20\% = \dfrac{1}{5}$ $7.00 | $7.00 | $7.00 | $7.00 | $7.00 |

or in other words,

$$1067 \div 1.36 = ?.$$

The answer is (approximately) 784.56.

(b)

$$1255 \div 843 = 1.489$$

(approximately). Multiplying by 100, we see that 1255 is 148.9% of 843.

4. From the information in the exercise,

$$(\frac{15}{100}) \times (\text{US budget}) = 241 \text{ billion}.$$

Therefore the US budget is

$$(\$241 \text{ billion}) \div .15 = \$1,606.7 \text{ billion},$$

which is about 1.6 trillion dollars.

5. Because

$$75\% = \frac{3}{4},$$

we can think of the \$12.3 million that make up 75% of the budget as distributed equally among the 3 parts of $\frac{3}{4}$, as shown in Figure 2.59. Each of those 3 parts must therefore contain \$4.1 million. The full budget is made of 4 of those parts. Therefore the full budget must be \$16.4 million.

6. As Figure 2.60 shows, 65% can be thought of as $50\% + 10\% + 5\%$. Now 50% of 4800 is $\frac{1}{2}$ of 4800, which is 2400. Since 10% of 4800 is $\frac{1}{10}$ of 4800, which is 480, therefore 5% is half of 480, which is 240. Therefore 65% of 4800 is $2400 + 480 + 240$, which is 3120, so 3120 gallons of water are left in the tank.

7. As shown in Figure 2.61, if a rectangle representing a full dose of 20 grams of medicine is divided into 10 equal parts, then each of those parts represents 10% of a full dose of medicine. Each part also must represent 2 grams of medicine. 9 grams of medicine is therefore represented by 4 full parts and half of a 5th part. Therefore 9 grams of medicine is 45%.

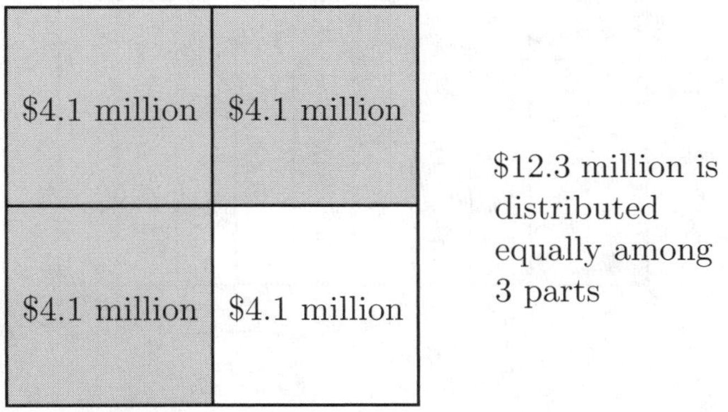

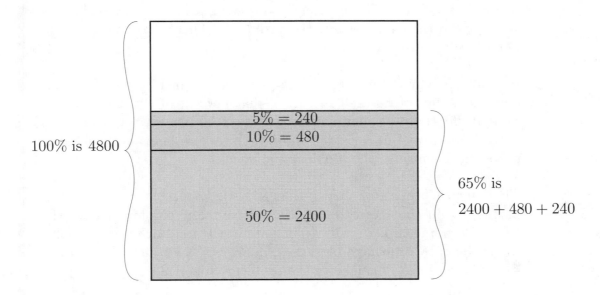

Figure 2.59: Calculating the Full Budget

Figure 2.60: Calculating 65%

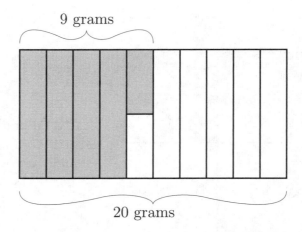

Figure 2.61: What Percent?

Problems for Section 2.7 on Percent

1. If 36,000 people is 15% of a population, then what is the total population? First solve the problem by drawing a picture. Explain how your picture helps you solve the problem. Then solve the problem numerically.

2. In Happy Valley the average rainfall in July is 5 inches, but this year, only 3.5 inches of rain fell in July. What percent of the average July rainfall did Happy Valley receive this year? First solve the problem by drawing a picture. Explain how your picture helps you solve the problem. Then solve the problem numerically.

3. The Biggo Corporation hopes that 95% of its 4,600 employees will participate in the charity fund-drive. How many employees does the Biggo Corporation hope will participate in the fund-drive? First solve the problem by drawing a picture. Explain how your picture helps you solve the problem. Then solve the problem numerically.

4. Malcolm ran 75% as far as Susan. How far did Susan run as a percentage of Malcolm's running distance? Draw a picture or diagram to help you solve the problem. Use your picture to help you explain your answer.

5. GrandMart sells 115% as much soda as BigMart. How much soda does BigMart sell, calculated as a percentage of GrandMart's soda sales?

6. Connie and Benton bought identical plane tickets, but Benton spent more than Connie (and Connie did not get her ticket for free).

    (a) If Connie spent 75% as much as Benton, then did Benton spend 125% as much as Connie? If not, then what percentage of Connie's ticket's price did Benton spend?

    (b) If Benton spent 125% as much as Connie, then did Connie spend 75% as much as Benton? If not, then what percentage of Benton's ticket's price did Connie spend?

7. 75% of the items sold at a newstand are newspapers. Does this mean that 75% of the newstand's income comes from selling newspapers? Why or why not? Write a paragraph discussing this. Include examples to illustrate your points.

8. There are two flasks. One contains water and the other contains an equal amount of wine (more than one cup). One cup of the wine is poured into the water flask and mixed thoroughly. Then one cup of this water/wine mixture in the water flask is poured into the wine flask and mixed thoroughly.

    (a) Without doing any calculations, which do you think will be greater, the percentage of wine in the water flask or the percentage of water in the wine flask? Explain your reasoning.

    (b) Now use calculations to figure out the answer to the problem in part (a). Work with specific quantities of water and wine (pick an amount, but remember that both flasks have the same amount at the start, and both contain more than one cup) and calculate the quantities and the percentages. Do you get a different answer than you got in part (a)? If so, reconcile your answers.

2.8   Comparing Sizes of Decimals

Given two decimals, they are either equal, or one is greater than the other. We will compare decimals from two different points of view. First we will

consider decimals as representing amounts in order to compare them. Then
we will use number lines to compare decimals.

Comparing Decimals by Viewing Them as Amounts

If A and B are decimals that are not negative, then we say that A is **greater
than** B, and we write

$$A > B,$$

if A represents a larger quantity than B. Similarly, we say that A is **less
than** B, and we write

$$A < B,$$

if A represents a smaller quantity than B. A good way to remember which
symbol to use is to notice that the "wide side" faces the larger number. Some
teachers have their students draw alligator teeth on the symbols to help them:
alligators are hungry and always want to eat the larger amount.

$$2 \lessdot 5$$

Figure 2.62: Alligator Teeth on a *Less Than* Symbol

For example,

$$.154 < .2$$

because

$$
\begin{aligned}
.154 &= 1 \cdot \frac{1}{10} + 5 \cdot \frac{1}{100} + 4 \cdot \frac{1}{1000}, \\
.2 &= 2 \cdot \frac{1}{10},
\end{aligned}
$$

and, as you can see in Figure 2.63, the combined amount of 5 hundredths
and 4 thousandths make less than 1 tenth. In fact, extrapolating from Figure 2.63, you can probably tell that even .199, which has 9 hundredths
and 9 thousandths—the largest possible amount of each in a decimal—
is still less than .2. Even if there were values in the ten-thousdandths,

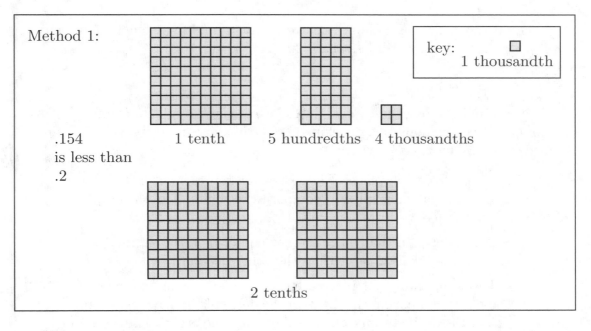

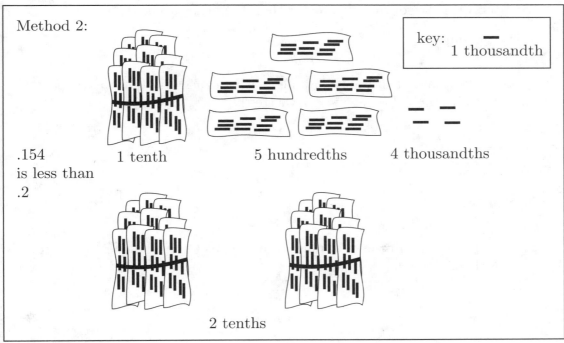

Figure 2.63: .154 Is Less Than .2

hundred-thousandths, millionths places, and so on, any decimal of the form
$.1****\ldots****$ is still less than .2,

$$.1****\ldots**** < .2.$$

By thinking along these lines, the following general rule for comparing two non-negative decimals was developed (this rule has one technical exception, which will be discussed below):

> Starting from the place of largest value represented in the decimals (the left-most place), compare the digits of both decimals in that place.
>
> —If the digits in that place are not equal, then the decimal with the larger digit is greater than the other decimal.
>
> —If the digits in that place are equal, keep moving one place to the right, comparing the digits of the decimals in like places, until one of the decimals has a larger digit than the other. The decimal with this larger digit is greater than the other decimal.

For example,

$$1234 > 789$$

because 1234 has a 1 in the thousands place and 789 has 0 in the thousands place (even though this zero is not written).

$$1234$$
$$\updownarrow$$
$$0789$$

Similarly

$$.89 < 1.2$$

because 1.2 has a 1 in the ones place, but .89 has a 0 in the ones place (even though this zero is not written).

$$0.89$$
$$\updownarrow$$
$$1.2$$

Finally,

$$1.2378 < 1.24$$

because both decimals have a 1 in the ones place and a 2 in the tenths place, but 1.24 has a larger digit, namely 4, in the hundredths place than does 1.2378, which only has a 3.

$$1.2378$$
$$\updownarrow \ \updownarrow\updownarrow$$
$$1.24$$

Notice that determining which decimal is greater is similar to putting words in alphabetical order, except that for decimals, we compare values in *like places*, whereas for words, we compare letters starting from the *beginning* of each word. Maybe this is why a decimal such as 1.01 might at first glance look like it is less than .998.

A Technical Exception to the Rule for Comparing Decimals

There is a rare exception to the method above for determining which of two decimals is greater. The exception occurs when one of the decimals has an infinitely repeating 9, such as

$$37.569999999999\ldots,$$

where the 9s repeat forever. In this case, the method above would lead us to say that

$$37.57 > 37.569999999999\ldots,$$

however, this is *not true*. In fact,

$$37.57 = 37.569999999999\ldots.$$

This topic will be explored in greater detail in Chapter 6. Meanwhile, Figure 2.64 may make it plausible to you that

$$.99999\ldots = 1.$$

Using Number Lines to Compare Decimals

It is especially easy to compare decimals when they are plotted on a number line. In addition, negative decimals can be compared along with positive ones.

If A and B are decimals plotted on a number line, then:

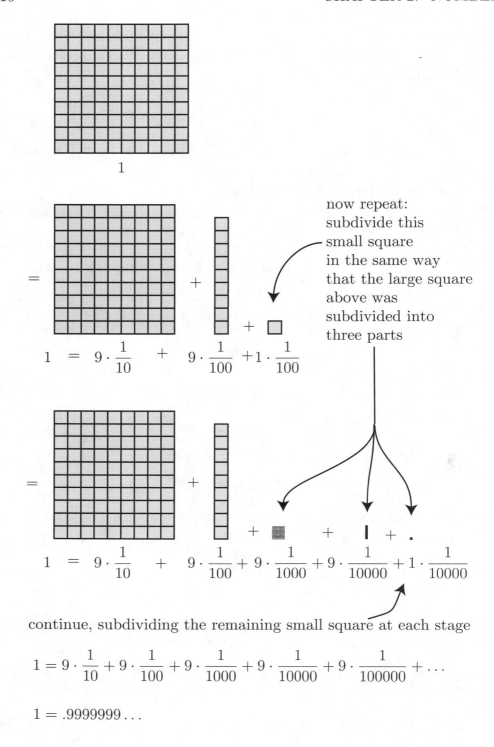

Figure 2.64: Showing That $1 = .9999\ldots$

$A < B$ provided that A is to the *left* of B on the number line,

$A > B$ provided that A is to the *right* of B on the number line.

These interpretations of *greater than* and *less than* are consistent with the ones given in the previous section for positive decimals, which were viewed as representing quantities. This is because if a positive decimal A represents a larger quantity than another positive decimal B, then A will be plotted farther from 0 than B. Since both are plotted to the right of 0, therefore A will be to the right of B. Likewise if A represents a smaller quantity than B.

With the number line interpretation of $>$ and $<$ it is easy to compare negative decimals. For example, since we plot -2.7 to the *left* of -1.5, as seen in Figure 2.65 therefore

$$-2.7 < -1.5.$$

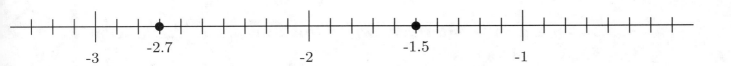

Figure 2.65: -2.7 Is Less Than -1.5

Comparing Negative Decimals by Viewing Them as *Owed* Amounts

In addition to explaining why $-2.7 < -1.5$ by using a number line, we can also explain why $-2.7 < -1.5$ by considering these negative decimals as owed amounts. -2.7 is represented by *owing* \$2.70 and -1.5 is represented by *owing* \$1.50, in other words, to *have* $-\$2.70$ is to *owe* \$2.70, and to *have* $-\$1.50$ is to *owe* \$1.50. If you owe \$2.70 you *have less* than if you owe \$1.50, so therefore

$$-2.7 < -1.5.$$

In general, the more you owe, the less you have, and the less you owe, the more you have, or in other words:

$$\text{if } A > B \text{ then } -A < -B$$

$$\text{if } A < B \text{ then } -A > -B.$$

For example, since

$$7 > 4,$$

therefore

$$-7 < -4.$$

Notice that the minus signs cause the $>$ symbol to reverse.

Class Activity 2O: Finding Smaller and Smaller Decimals

Class Activity 2P: Finding Decimals Between Decimals

Class Activity 2Q: Decimals Between Decimals on Number Lines

Class Activity 2R: *Greater Than* and *Less Than* with Negative Decimals

Exercises for Section 2.8

1. What's wrong with a grocery store sign that says "squash, .99 cents per pound"?

2. Which is greater, .01000001 or .0099999999999?

3. Draw a picture like Figure 2.63 demonstrating that $1.2 > .89$.

4. Children who have heard of a google will often think it's the largest number. Is it? What about a googleplex—is that the largest number? Is there a largest number?

5. Find a decimal between $\sqrt{2}$ and 1.41422 and plot all three numbers on a number line.

6. Use a number line to show that $1.2 > .89$.

7. Use a number line to show that $-1.2 < -.89$.

8. Explain why $-1.2 < -.89$ without using a number line.

9. The smallest integer that is greater than zero is 1. Is there a smallest *real number* that is greater than zero?

Answers to Exercises for Section 2.8

1. .99 cents is less than one cent! It's hard to imagine squash, or any other vegetable, selling for less than 1 cent per pound. The store probably means that the price is $.99 per pound, or in other words, 99 cents per pound.

2. .01000001 is greater because of the 1 in the hundredths place.

3. See Figure 2.66.

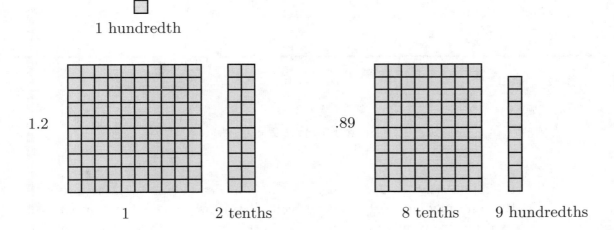

Figure 2.66: 1.2 > .89

4. Notice that a googolplex is greater than a googol. The number googolplex plus one, for example, is greater than a googolplex. There is no largest number because no matter what number you choose as candidate for the largest number, that number plus one is a larger number.

5. $\sqrt{2} = 1.414213\ldots$, so 1.414215 is one example of a decimal between $\sqrt{2}$ and 1.41422. Of course there are many others. See Figure 2.67

6. See Figure 2.68.

Figure 2.67: Plotting Three Numbers

Figure 2.68: 1.2 > .89

7. See Figure 2.69.

Figure 2.69: −1.2 < −.89

8. To explain why −1.2 < −.89, we can consider having −1.2 as owing $1.20 and having −.89 as owing $.89. If you owe $1.20 then you *have less* than if you only owe $.89. Therefore having −1.2 means you have less than if you have −.89, so that −1.2 < −.89.

9. No, there is no smallest real number that is greater than 0. Every real number can be expressed as a decimal. Consider the list of decimals:

$$.1$$
$$.01$$
$$.001$$
$$.0001$$
$$.00001$$
$$.000001$$
$$\vdots$$

The decimals in this list get smaller and smaller, getting ever closer to 0 without ever reaching 0. No matter what decimal you choose that is greater than 0, the decimals in the list above will eventually get smaller than your chosen decimal. Therefore there can be no smallest positive decimal, and hence no smallest positive real number.

Problems for Section 2.8

1. Mary says that the numbers in this list are getting bigger:

$$4.1, 4.2, 4.3, 4.4, 4.5, 4.6, 4.7, 4.8, 4.9, 4.10, 4.11, 4.12, 4.13,$$
$$4.14, 4.15, \ldots, 4.18, 4.19, 4.20, 4.21, 4.22, \ldots, 4.29, 4.30,$$
$$4.31, \ldots 4.98, 4.99, 4.100, 4.101, \ldots.$$

Is Mary right or not? If she's not right, then which numbers on the list are immediately to the left of a smaller number?

2. Mark says that .178 is greater than .25. Describe several ways to convince him that it's not.

3. Explain why $-3.25 < -1.4$ in two different ways.

4. Describe an infinite list of decimals, all of which are greater than 3.514, but that get closer and closer to 3.514.

2.9 Comparing Sizes of Fractions

Given two decimals, we can determine which one is greater by comparing the digits of the decimals. But the situation for fractions is more complicated. This is because every fraction is equal to infinitely many other fractions, so unlike whole numbers, we can't always tell just by comparing digits of fractions whether or not they are equal, or whether one is greater than the other. Even though you will probably recognize that a familiar fraction such as $\frac{1}{2}$ is equal to $\frac{3}{6}$ or $\frac{50}{100}$, can you tell right away, just by looking, that

$$\frac{411}{885} \quad \text{and} \quad \frac{548}{1180}$$

are equal? It is not obvious.

There are three standard methods for determining whether two fractions are equal, or if not, which one is greater.

Comparing Fractions by Converting to Decimals

Every fraction can be converted to a decimal by dividing the denominator into the numerator. Therefore two fractions can be compared simply by converting them both to decimals and comparing the decimals.

Is $\frac{17}{35}$ equal to $\frac{43}{87}$, or if not, which is greater?

$$\frac{17}{35} = 17 \div 35 = .4857\ldots,$$
$$\frac{43}{87} = 43 \div 87 = .4942\ldots.$$

Since

$$.4942\ldots > .4857\ldots,$$

therefore

$$\frac{43}{87} > \frac{17}{35}.$$

Comparing Fractions by Using Common Denominators

When two fractions have the same denominator, then you *can* determine whether they are equal, or if not, which is greater, just by looking at them. The fractions

$$\frac{784}{953} \text{ and } \frac{621}{953},$$

have the same denominator. Since

$$784 > 621,$$

therefore

$$\frac{784}{953} > \frac{621}{953}.$$

In general, *if two fractions have the same denominator, then the one with the greater numerator is greater; if the numerators are equal, then the fractions are equal.*

Why does this rule for comparing fractions make sense? If you have two identical pies, say, and you divide each pie into the same number of equal pieces (same denominator), then you will eat more of one pie if you eat more

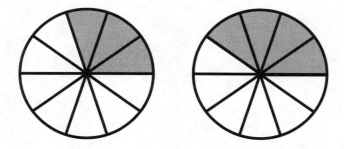

Figure 2.70: Comparing Fractions with Equal Denominators

pieces of that pie (larger numerator). See Figure 2.70. If you eat the same number of pieces of each pie, then you will have eaten the same amount of each pie.

But what if you have two fractions and they *don't* have the same denominator? How can you tell whether they are equal, or if not, which fraction is greater? You can give the two fractions a common denominator. in other words, the *same* denominator. It can be *any* common denominator—it *does not* have to be the least one. A common denominator that *always works* is obtained by multiplying the two denominators. Consider the two fractions above:

$$\frac{411}{885} \quad \text{and} \quad \frac{548}{1180}.$$

Both can be given the common denominator

$$885 \times 1180.$$

Then

$$\frac{411}{885} = \frac{411 \cdot 1180}{885 \cdot 1180} = \frac{484980}{1044300}$$

and

$$\frac{548}{1180} = \frac{548 \cdot 885}{1180 \cdot 885} = \frac{484980}{1044300}.$$

The two fractions $\frac{411}{885}$ and $\frac{548}{1180}$ are equal because when both are written with the denominator $885 \cdot 1180 = 1044300$, they have the same numerator. See Figure 2.71 for another example.

Which is greater,

$$\frac{4}{9} \quad \text{or} \quad \frac{3}{5}?$$

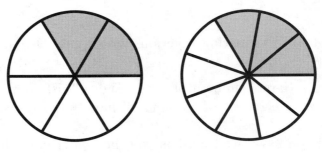

Are these fractions equal?

Subdivide each into like parts:

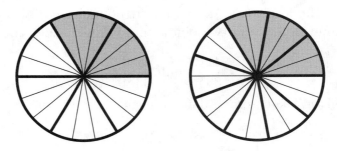

Each is 6/16, so they are equal.

Figure 2.71: Seeing That Two Fractions Are Equal

We can give both fractions the common denominator of $9 \cdot 5 = 45$:

$$\frac{4}{9} = \frac{4 \cdot 5}{9 \cdot 5} = \frac{20}{45}$$

$$\frac{3}{5} = \frac{3 \cdot 9}{5 \cdot 9} = \frac{27}{45},$$

as shown in Figure 2.72. Since the denominators are now the same, and the

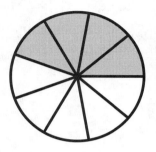 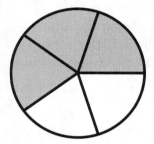

Which fraction is bigger?

Subdivide into like parts:

4×5 parts versus 3×9 parts

Figure 2.72: Comparing $\frac{4}{9}$ and $\frac{3}{5}$

numerator 27 is greater than the numerator 20, therefore

$$\frac{27}{45} > \frac{20}{45},$$

so that

$$\frac{3}{5} > \frac{4}{9}.$$

Comparing Fractions by *Cross-Multiplying*

Notice that above, to show that the two fractions $\frac{411}{885}$ and $\frac{548}{1180}$ are equal, all we really had to do was check that

$$411 \cdot 1180 = 548 \cdot 885,$$

because $411 \cdot 1180$ and $548 \cdot 885$ were the numerators when we gave $\frac{411}{885}$ and $\frac{548}{1180}$ a common denominator. So, all we really had to do was to multiply the numerator of each fraction $\frac{411}{885}$ and $\frac{548}{1180}$ by the denominator of the other fraction, and check if those two resulting numbers are equal.

Similarly, above, to show that

$$\frac{3}{5} > \frac{4}{9},$$

all we really had to do was check that

$$3 \cdot 9 > 4 \cdot 5,$$

because $3 \cdot 9$ and $4 \cdot 5$ were the numerators when we gave $\frac{4}{9}$ and $\frac{3}{5}$ the common denominator of $9 \cdot 5$.

Generalizing the above, if $\frac{A}{B}$ and $\frac{C}{D}$ are two fractions of whole numbers (where neither B nor D is zero), then we can give these fractions a common denominator:

$$\frac{A}{B} = \frac{A \cdot D}{B \cdot D},$$
$$\frac{C}{D} = \frac{C \cdot B}{D \cdot B}.$$

To compare $\frac{A}{B}$ and $\frac{C}{D}$, we can compare

$$\frac{A \cdot D}{B \cdot D}$$

and

$$\frac{C \cdot B}{D \cdot B}$$

instead. Because the denominators $D \cdot B$ and $B \cdot D$ are equal, therefore we only need to compare the numerators

$$A \cdot D \quad \text{and} \quad C \cdot B.$$

We conclude that

- the fractions $\frac{A}{B}$ and $\frac{C}{D}$ are equal exactly when $A \cdot D$ and $C \cdot B$ are equal;

- $\frac{A}{B}$ is greater than $\frac{C}{D}$ exactly when $A \cdot D$ is greater than $C \cdot B$;

- $\frac{A}{B}$ is less than $\frac{C}{D}$ exactly when $A \cdot D$ is less than $C \cdot B$.

Stated more symbolically:

$$\frac{A}{B} = \frac{C}{D} \text{ exactly when } A \cdot D = C \cdot B;$$

$$\frac{A}{B} > \frac{C}{D} \text{ exactly when } A \cdot D > C \cdot B;$$

$$\frac{A}{B} < \frac{C}{D} \text{ exactly when } A \cdot D < C \cdot B.$$

This method for checking if two fractions are equal, or if not, which one is greater, is often called the **cross-multiplying** method. *Notice that it is just a way to check if the numerators are equal when the two fractions are given the common denominator obtained by multiplying the two original denominators.*

Using Other Reasoning To Compare Fractions

The standard methods described above for comparing fractions are useful because they are efficient and they always work. But sometimes it is possible to use other reasoning to compare fractions. Why would we want to use other reasoning when we already have several good methods? In some cases, other reasoning can be more efficient. But even more importantly, when we have to reason in other ways, it forces us to think about the meaning of fractions. If we only use efficient methods that we don't have to think much about, then we tend to lose sight of the meaning of what we are doing, and we may start to see mathematics as mechanistic and meaningless.

Consider the case of determining whether $\frac{3}{5}$ and $\frac{4}{9}$ are equal, and if not, which is larger. We can reason that $\frac{4}{9}$ is less than $\frac{1}{2}$,

$$\frac{4}{9} < \frac{1}{2},$$

because if we divide an object into 9 equal pieces, then it would take four and a half pieces to make half of the object, so 4 pieces is certainly less than half; similarly, we can say that

$$\frac{1}{2} < \frac{3}{5}$$

because if we divide an object into 5 equal pieces, then two and a half of them would make half the object, so three pieces is more than half the object. Since $\frac{4}{9}$ is less than a half, and $\frac{3}{5}$ is more than a half, therefore $\frac{3}{5}$ is the larger fraction.

Which fraction is greater,

$$\frac{5}{8} \text{ or } \frac{5}{9}?$$

Both fractions represent 5 parts, but $\frac{5}{8}$ is 5 parts when the whole is divided into 8 parts, whereas $\frac{5}{9}$ is 5 parts when the whole is divided into 9 parts. But if an object is divided into 8 equal parts then each part is smaller than if the object is divided into 9 equal parts—more parts making up the same whole means each part has to be smaller. So 5 of 8 equal parts must be greater than 5 of 9 equal parts. Thus

$$\frac{5}{8} > \frac{5}{9},$$

as shown in Figure 2.73.

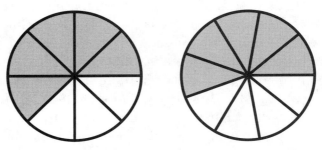

Both have 5 parts, but 8ths are bigger than 9ths.

Figure 2.73: Comparing $\frac{5}{8}$ and $\frac{5}{9}$

Class Activity 2S: Can We Compare Fractions This Way?

Class Activity 2T: Comparing Fractions by Reasoning

Class Activity 2U: Rules For Comparing Fractions

Exercises for Section 2.9 on Comparing Sizes of Fractions

1. Compare the sizes of the following pairs of fractions in two ways: by "cross-multiplying" and by giving the fractions common denominators. How are these methods related?

$$\frac{2}{3} \text{ and } \frac{3}{5}$$

$$\frac{8}{13} \text{ and } \frac{13}{21}$$

$$\frac{15}{20} \text{ and } \frac{6}{8}$$

$$\frac{1}{4} \text{ and } \frac{3}{8}$$

$$\frac{4}{14} \text{ and } \frac{5}{15}$$

2. Complete the following to make true statements about comparing fractions:

   (a) If two fractions have the same denominators, then the one with .... is greater.

   (b) If two fractions have the same numerator, then the one with .... is greater.

   Use the meaning of fractions to explain why your answers make sense.

3. For each of the following pairs of fractions, determine which is larger. Use reasoning *other than* finding common denominators, cross-multiplying, or converting to decimals in order to explain your answer.

   (a) $\frac{6}{11}$ versus $\frac{6}{13}$

   (b) $\frac{5}{8}$ versus $\frac{7}{12}$

   (c) $\frac{5}{21}$ versus $\frac{7}{24}$

(d) $\frac{21}{22}$ versus $\frac{56}{57}$

(e) $\frac{97}{100}$ versus $\frac{35}{38}$

4. Explain why

$$\frac{A}{B} > \frac{C}{D} \quad \text{exactly when} \quad A \cdot D > C \cdot B.$$

Answers to Exercises for Section 2.9 on Comparing Sizes of Fractions

1. See the text for the relationship between the *cross-multiplying* and *common denominator* methods for comparing fractions.

Notice that to compare $\frac{15}{20}$ and $\frac{6}{8}$ with the common denominator method you could choose a smaller common denominator than $20 \cdot 8$. For instance, you might choose 40 as common denominator. In this case, you compare the numerators $2 \cdot 15$ and $5 \cdot 6$ rather than comparing $8 \cdot 15$ and $20 \cdot 6$, which is what you compare with the cross-multiplying method. Similarly with $\frac{1}{4}$ and $\frac{3}{8}$. Of course it's easiest just to use the common denominator 8 in that case.

fractions	common denominator	cross multiplying	conclusion
$\frac{2}{3}$ $\frac{3}{5}$	$\frac{2\cdot5}{3\cdot5} > \frac{3\cdot3}{5\cdot3}$	$2 \cdot 5 > 3 \cdot 3$	$\frac{2}{3} > \frac{3}{5}$
$\frac{8}{13}$ $\frac{13}{21}$	$\frac{8\cdot21}{13\cdot21} < \frac{13\cdot13}{21\cdot13}$	$8 \cdot 21 < 13 \cdot 13$	$\frac{8}{13} < \frac{13}{21}$
$\frac{15}{20}$ $\frac{6}{8}$	$\frac{15\cdot8}{20\cdot8} = \frac{6\cdot20}{8\cdot20}$	$15 \cdot 8 = 6 \cdot 20$	$\frac{15}{20} = \frac{6}{8}$
$\frac{1}{4}$ $\frac{3}{8}$	$\frac{1\cdot8}{4\cdot8} < \frac{3\cdot4}{8\cdot4}$	$1 \cdot 8 < 3 \cdot 4$	$\frac{1}{4} < \frac{3}{8}$
$\frac{4}{14}$ $\frac{5}{15}$	$\frac{4\cdot15}{14\cdot15} < \frac{5\cdot14}{15\cdot14}$	$4 \cdot 15 < 5 \cdot 14$	$\frac{4}{14} < \frac{5}{15}$

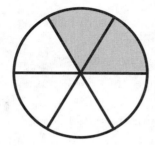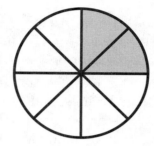

Figure 2.74: Two Pieces Eaten in Each but Different Size Pie Pieces

2. (a) If two fractions have the same denominators, then the one with the greater numerator is greater. See the text for an explanation why.

   (b) If two fractions have the same numerator, then the one with the smaller denominator is greater. Say you have two identical pies, and you divide one pie into 6 equal pieces and the other into 8 equal pieces as in Figure 2.74. If you eat two pieces of each pie, then, even though you've eaten the same number of piece from each pie, you have eaten more of the 6-piece pie because the pie pieces in the 6-piece pie are bigger than the pie pieces in the 8-piece pie—fewer pieces making up the same size pie means each piece must be bigger. The reasoning works the same way when other counting numbers replace 6 and 8 and the second number is greater than the first.

3. (a) $\frac{6}{11} > \frac{6}{13}$ If you divide a pie into 13 pieces then each piece will be smaller than if you divide that same pie into 11 pieces. So 6 pieces of a pie divided into 11 pieces will be more than 6 pieces of a pie divided into 13 pieces.

   (b) $\frac{5}{8}$ is one piece more than half $\left(\frac{4}{8}\right)$ and $\frac{7}{12}$ is one piece more than half $\left(\frac{6}{12}\right.$. But when a pie is divided into 8 pieces, each piece is larger than when an identical pie is divided into 12 pieces—fewer pieces making up the same amount means each piece must be larger. Since the eighths pieces are larger than the twelfths pieces, and since each fraction, $\frac{5}{8}$ and $\frac{7}{12}$ is one piece more than one half, therefore $\frac{5}{8} > \frac{7}{12}$.

   (c) Notice that both fractions are close to $\frac{1}{4}$. Since $\frac{1}{4} = \frac{5}{20}$, therefore

$\frac{5}{21}$ is a little less than $\frac{1}{4}$ (see the reasoning of the previous part's answer). Since $\frac{1}{4} = \frac{6}{24}$, therefore $\frac{7}{24}$ is a little bigger than $\frac{1}{4}$ (see the reasoning of previous part's answer). So $\frac{5}{21} < \frac{7}{24}$.

(d) Notice that each fraction is "one piece less than 1 whole." $\frac{21}{22}$ is $\frac{1}{22}$ less than a whole, and $\frac{56}{57}$ is $\frac{1}{57}$ less than a whole. But $\frac{1}{22}$ is bigger than $\frac{1}{57}$ because if you divide a pie into 22 pieces, each piece will be bigger than if you divide an identical pie into 57 pieces. Therefore $\frac{21}{22} < \frac{56}{57}$ because $\frac{21}{22}$ is a bigger piece away from a whole than is $\frac{56}{57}$.

(e) The same reasoning used in the answer to the previous part applies here too. This time, each fraction is 3 pieces away from 1 whole.

4. See text.

Problems for Section 2.9 on Comparing Sizes of Fractions

1. Julie says that the picture in Figure 2.75 shows that

$$\frac{1}{4} > \frac{1}{2}$$

by comparing the areas. Explain carefully what is wrong with Julie's reasoning.

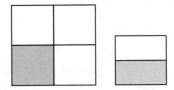

Figure 2.75: $\frac{1}{4} > \frac{1}{2}$?

2. Give *two different methods* for solving the following problem. Find three different fractions in between $\frac{3}{5}$ and $\frac{2}{3}$ whose numerators and denominators are all whole numbers.

3. A student says that $\frac{1}{5}$ is halfway between $\frac{1}{4}$ and $\frac{1}{6}$. Use a *carefully drawn* number line to show that this is not correct. What fraction *is* halfway between $\frac{1}{4}$ and $\frac{1}{6}$? Explain your reasoning.

4. You may combine your answers to all three parts of this problem.

    (a) Is it valid to compare

    $$\frac{30}{70} \quad \text{and} \quad \frac{20}{50}$$

    by "cancelling" the 0s and comparing

    $$\frac{3}{7} \quad \text{and} \quad \frac{2}{5}$$

    instead? Explain your answer.

    (b) Is it valid to compare

    $$\frac{15}{25} \quad \text{and} \quad \frac{105}{205}$$

    by "cancelling" the 5s and comparing

    $$\frac{1}{2} \quad \text{and} \quad \frac{10}{20}$$

    instead? Explain your answer.

    (c) Write a paragraph discussing the distinction between your answer in (a) and your answer in (b).

5. Which fraction is bigger, $\frac{15}{31}$ or $\frac{23}{47}$? Use the meaning of fractions and reasoning *other than* finding common denominators, cross-multiplying, or converting to decimals in order to explain your answer.

6. Minju says that fractions that use bigger numbers are greater than fractions that use smaller numbers. Make up two problems for Minju to help her reconsider her ideas. For each problem, explain how to solve it, and explain why you chose that problem for Minju.

7. Malcolm says that

    $$\frac{8}{11} > \frac{7}{10}$$

    because

    $$8 > 7 \quad \text{and} \quad 11 > 10.$$

    Even though it is true that $\frac{8}{11} > \frac{7}{10}$, is Malcolm's reasoning correct? If Malcolm's reasoning is correct, explain why clearly; if Malcolm's reasoning is not correct, give Malcolm two examples that show why not.

8. Consider the following list of fractions:

$$\frac{1}{1}, \quad \frac{2}{1}, \quad \frac{3}{2}, \quad \frac{5}{3}, \quad \frac{8}{5}, \quad \cdots$$

You do not have to explain your answers to the parts below.

(a) Describe a pattern in the list of fractions and use your description to find the next 5 entries in the list after $\frac{8}{5}$. You will now have the first 10 entries in the list of fractions.

(b) Use either the *cross-multiplying* method or the *common denominator* method to compare the sizes of the 1st, 3rd, 5th, 7th, and 9th fractions in the list. Describe a pattern in the sizes of these fractions. Describe a pattern that occurs when you compare the fractions.

(c) Use either the *cross-multiplying* method or the *common denominator* method to compare the sizes of the 2nd, 4th, 6th, 8th, and 10th fractions in the list. Describe a pattern in the sizes of these fractions. Describe a pattern that occurs when you compare the fractions.

(d) Convert the 10 fractions on your list to decimals, and plot them on a number line. *Zoom in* on portions of your number line, as in Figure 2.19, so that you can show clearly where each decimal is plotted relative to the others.

(e) If you could find more and more entries in the list of fractions, and plot them on a number line, in what region of the number line would they be located? Do you think these numbers would get closer and closer to some one number?

9. Suppose you start with a proper fraction and you add 1 to the numerator and to the denominator. For example, if you started with $\frac{2}{3}$, then you'd get a new fraction $\frac{2+1}{3+1} = \frac{3}{4}$.

(a) Give at least 5 examples of proper fractions $\frac{A}{B}$. In each example, compare the sizes of $\frac{A}{B}$ and $\frac{A+1}{B+1}$. What do you notice? Make sure you are working with *proper* fractions (where the numerator is less than the denominator).

(b) Frank says that if $\frac{A}{B}$ is a proper fraction, then $\frac{A+1}{B+1}$ is always greater than $\frac{A}{B}$ because $\frac{A+1}{B+1}$ has more parts. Regardless of whether or not Frank's conclusion is correct, discuss whether or not Frank's *reasoning* is valid. Did Frank give a convincing explanation that $\frac{A+1}{B+1}$ is greater than $\frac{A}{B}$? If not, what objections could you make to Frank's reasoning?

(c) Give a careful and thorough explanation for the phenomenon you discovered in part (a): compare the sizes of $\frac{A}{B}$ and $\frac{A+1}{B+1}$ and explain why this is so.

2.10   Rounding Decimals

Decimals are often used to describe the size of actual quantities. The distance between two locations might be described as 2.7 miles, the population of a city might be given as 87,000, a tax rebate might be described as $1.3 trillion. In cases where a decimal describes an actual amount, a precise decimal representation is often either not required, or not available. In such cases, the actual decimal describing the amount has been *rounded*.

Specified Rounding

To **round** a decimal means to find a nearby decimal that has fewer (or not more) non-zero digits. *What qualifies as "nearby" depends entirely on the context*, so that in realistic situations, rounding generally requires judgement and common sense. We can indicate what we mean by "nearby" by stating the type of rounding we want: we can ask that a decimal be rounded to the nearest hundred, or to the nearest ten, or to the nearest one, or to the nearest tenth, or to the nearest hundredth, and so on. To round a given decimal to the nearest hundred, find the decimal closest to your given decimal that has only zeros in places smaller than the hundreds place. To round a given decimal to the nearest tenth, find the decimal closest to your given decimal that has only zeros in places smaller than the tenths place. Similarly for any other kind of specified rounding.

To understand specified rounding, it helps to consider a number line with tick marks chosen according to the specified type of rounding. When rounding to the nearest hundred, consider a number line with large tick marks spaced 100 apart, as in Figure 2.76. The decimals at the large tick marks

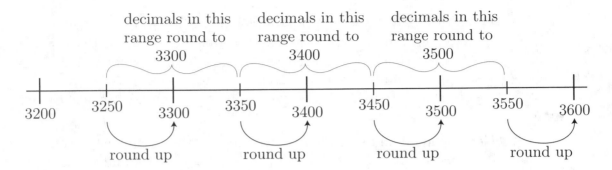

Figure 2.76: Rounding to the Nearest Hundred

are exactly those decimals that have only zeros in places smaller than the hundreds place. So when you round to the nearest hundred, you are finding the large tick mark that is closest to the location of the number you are rounding. Thus, to round 3367 to the nearest hundred, notice that 3367 lies between 3300 and 3400. But 3367 is closer to 3400 than to 3300 because 3367 is greater than 3350, which is half way between 3300 and 3400. Therefore round 3367 to 3400.

When rounding to the nearest tenth, consider a number line with large tick marks spaced $\frac{1}{10}$ apart, as in Figure 2.77. The decimals at these large

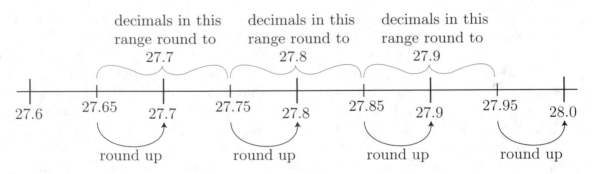

Figure 2.77: Rounding to the Nearest Tenth

tick marks are exactly those decimals that have only zeros in places smaller than the tenths place. So when you round to the nearest tenth, you are finding the large tick mark that is closest to the location of the number you are rounding. Thus, to round 27.849 to the nearest tenth, notice that 27.849

lies between 27.8 and 27.9. But 27.849 is closer to 27.8 than to 27.9 because 27.849 is less than 27.85, which is half way between 27.8 and 27.9. Therefore round 27.849 to 27.8.

What do we do about decimals that are exactly half way in between two tick marks? For example, how do we round 3450 when we are rounding to the nearest hundred? How do we round 27.45 when we are rounding to the nearest tenth? We need a convention to break ties like these. The most common convention, and the one you should use unless otherwise specified, is to *round up* when there is a tie. So when rounding 3450 to the nearest hundred, round to 3500, as shown in Figure 2.76. When rounding 27.85 to the nearest tenth, round to 27.9, as shown in Figure 2.77. Another convention for rounding, one that is often used by engineers, is to *round to the nearest even* when there is a tie. With this convention, when rounding to the nearest hundred, 3450 would be rounded to 3400, rather than 3500, because 4 is even. Similarly, when rounding to the nearest tenth, 27.85 would be rounded to 27.8, rather than 27.9, because 8 is even.

Working with Decimals that Represent Actual Quantities

When a decimal is used to describe the size of an actual quantity, such as a distance or a population, we generally assume that this number has been rounded. Furthermore, we assume that *the way a decimal is written indicates the rounding that has taken place*. For example, when the distance between two locations is described as 6.2 miles, we assume from the presence of a digit in the tenths place but no smaller place that the actual distance has been rounded to the nearest tenth of a mile. Therefore the actual distance could be anywhere between 6.15 and 6.25 miles (but less than 6.25 miles). Similarly, if the population of a city is described as $93,000$, we assume from the non-zero entry in the thousands place and the zeros in all lower places that the actual population has been rounded to the nearest thousands. Therefore the actual population could be anywhere between $92,500$ and $93,500$ (but less than $93,500$).

Because of rounding, there is a slight difference in the meaning of decimals in the abstract, and decimals when they are used to describe the size of an actual quantity. In the abstract,

$$8.00 = 8,$$

but when we write 8 for the weight of an object in kilograms it has a different meaning than when we write 8.00 for the weight of an object in kilograms: writing 8 indicates that the actual weight has been rounded to the nearest one, whereas writing 8.00 means that the actual weight has been rounded to the nearest hundredth. If the weight is reported as 8.00 kilograms, then the actual weight could be anywhere between 7.995 and 8.005 kilograms. But if the weight is reported as 8 kilograms, then the actual weight could be anywhere between 7.5 and 8.5 kilograms. Therefore, writing 8.00 conveys that the weight is known much more accurately than if the weight is reported as 8 kilograms.

When you work a problem that involves real or realistic quantities, you should round your answer so that it does not appear to be more accurate than it actually is. Your answer cannot be any more accurate than the numbers you started with, so you should round your answer to fit with with the rounding of your initial numbers. For example, suppose that the population of a city is given as 1.6 million people, and that after some calculations, you project that the city will have 1.95039107199 million people in 10 years. Although the decimal 1.95039107199 may be the exact answer to your calculations, you should not report your answer this way, because it makes your answer appear to be more accurate than it actually is. We must assume that the initial number 1.6 is rounded to the nearest tenth. Therefore we should also round the answer, 1.95039107199, to the nearest tenth and report the projected population in 10 years as 2.0 million.

Class Activity 2V: Explaining Why a Procedure for Rounding Makes Sense

Class Activity 2W: Can We Round This Way?

Exercises for Section 2.10

1. Round 6.248 to the nearest tenth. Explain in words why you round the decimal the way you do. Use a number line to support your explanation.

2. Round 173.465 to the nearest hundred, to the nearest ten, to the nearest one, to the nearest tenth, and to the nearest hundredth.

3. The distance between two cities is described as 1500 miles. Should you assume that this is the exact distance between the cities? If not, what

can you say about the exact distance between the cities?

Answers to Exercises for Section 2.10

1. 6.248 is between 6.2 and 6.3. But because of the 4 in the hundredths place, 6.248 is less than 6.25, which is half way between 6.2 and 6.3. Therefore 6.248 is closer to 6.2 than to 6.3, and so 6.248 rounded to the nearest tenth is 6.2.

   The number line in Figure 2.78 shows what was stated above.

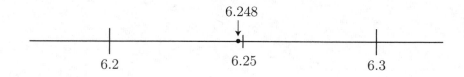

Figure 2.78: Rounding 6.248 to the Nearest Tenth

2. To the nearest hundred: 200

   To the nearest ten: 170

   To the nearest one: 173

   To the nearest tenth: 173.5

   To the nearest hundredth: 173.47

3. Because the reported distance has zeros in the tens and ones places, you should assume that the distance has been rounded to the nearest hundred. Therefore the exact distance between the two cities is probably not 1500 miles; it could be anywhere between 1450 miles and 1550 (but less than 1550).

Problems for Section 2.10

1. Round 2.1349 to the nearest hundredth. Explain in words why you round the decimal the way you do. Use a number line to support your explanation.

2. Round 9995.2 to the nearest ten. Explain in words why you round the decimal the way you do. Use a number line to support your explanation.

3. The weight of an object is reported as 10,000 pounds. Should you assume that this is the exact weight of the object? If not, what can you say about the exact weight of the object?

Chapter 3

Addition and Subtraction

This chapter is about the interpretations of addition and subtraction and about why the standard procedures for carrying out addition and subtraction problems make sense based on these interpretations. We will work with whole numbers, negative numbers, decimals, fractions, and even percents.

3.1   Interpretations of Addition and Subtraction

The most familiar ways of thinking about addition and subtraction are as *combining* and *taking away*. These hardly need explanation. But as we'll see in this section, we can also consider addition and subtraction on number lines, and this allows us to make sense of addition and subtraction involving negative numbers. Subtraction is also often used to *compare* numbers. This point of view leads to a common practice of shopkeepers for making change.

Addition and Subtraction as Combining and Taking Away

Typically, if A and B are two numbers, then the **sum**

$$A + B$$

represents the total number of objects you will have if you start with A objects and then get B more objects. The numbers A and B in a sum are called **terms**, **addends**, or **summands**.

You can represent the sum

$$149 + 85$$

as the total number of toothpicks you will have if you start with 149 toothpicks and you get 85 more toothpicks. You can represent the sum

$$34.5 + 7.89$$

as the total number of dollars you will have if you start with $34.5 = 34.50$ dollars and then get 7.89 more dollars. You can represent the sum

$$\frac{3}{4} + \frac{2}{3}$$

as the total number of cups of flour in a batch of dough if you start with $\frac{3}{4}$ of a cup of flour in the dough and add in another $\frac{2}{3}$ of a cup of flour.

Typically, if A and B are two numbers, the **difference**

$$A - B$$

represents the total number of objects you will have if you start with A objects and take away B of those objects. The numbers A and B in a difference can be called **terms**. The number A is sometimes called the **minuend** and the number B is sometimes called the **subtrahend**.

You can represent

$$142 - 83$$

as the number of toothpicks you have left if you start with 142 toothpicks and give away 83 toothpicks. You can represent

$$319 - 148.2$$

as the number of dollars you have left if you start with $319 and spend $148.20. You can represent

$$\frac{3}{4} - \frac{1}{8}$$

as the number of acres of land you have left if you start with $\frac{3}{4}$ of an acre and sell $\frac{1}{8}$ of an acre.

The interpretations of addition and subtraction given above are the most basic and simple interpretations, but notice that they don't really make sense for certain kinds of numbers. For example, the meaning of sums and differences like

$$\sqrt{2} + \sqrt{3},$$

$$2 + (-3),$$

and

$$1 - 5$$

seem murky with these interpretation. For example, how can you "take away" 5 objects when you only have 1? One way to interpret this is in terms of number of objects *owed*, but another way is to use number lines, as we will now see.

Addition And Subtraction on Number Lines

Addition and subtraction of all numbers, whether they are positive or negative negative, and whether they are represented by decimals or by fractions, can be interpreted easily using number lines.

If A and B are any two numbers, then they are represented by points on a number line. The sum

$$A + B$$

corresponds to the point on the number line that is located as follows:

- Start at A on the number line;

- move a distance equal to the number of units that B is away from 0:

  - move to the right if B is positive,

  - move to the left if B is negative.

- The resulting point is the location of $A + B$.

Where is $2 + (-3)$ on a number line? As shown in Figure 3.1), start at 2 and move 3 units to the left—move left because -3 is negative. You end up at -1, so

$$2 + (-3) = -1.$$

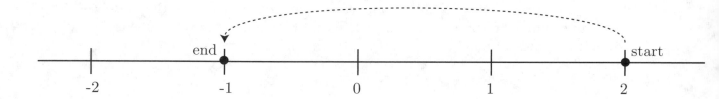

Figure 3.1: $2 + (-3) = -1$

Similarly, if A and B are any two numbers, the difference

$$A - B$$

corresponds to the point on the number line that is located as follows:

- Start at A on the number line;

- move a distance equal to the number of units that B is away from 0:

  - move to the *left* if B is positive,
  - move to the *right* if B is negative.

- The resulting point is the location of $A - B$.

Where is $(-2) - (-4)$ on a number line? As shown in Figure 3.2, start at -2 and move 4 units to the *right*—move right because -4 is negative and we are subtracting. You end up at 2, so $(-2) - (-4) = 2$.

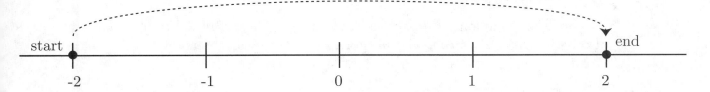

Figure 3.2: $(-2) - (-4) = 2$

The number line interpretations of addition and subtraction agree with the basic *combining* and *taking away* interpretations described above. The advantage of the the number line interpretations is that they apply to *all* numbers.

Relating Addition and Subtraction

Every statement about subtraction corresponds to a statement about addition. Namely, to say that
$$A - B = C,$$
is equivalent to saying that
$$A = C + B.$$

We can see why these two statements are equivalent by showing them on a number line, as seen in Figure 3.3.

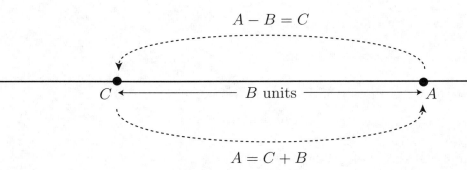

$$A - B = C$$

$$A = C + B$$

Figure 3.3: A Subtraction Statement Corresponds to an Addition Statement

Subtraction as Comparison

Although we commonly think of subtraction as *taking away*, subtraction can also be interpreted as *comparison*. For example, we can interpret

$$7 - 5$$

as the number of blocks that remain from 7 blocks when 5 blocks are taken away. But as seen in Figure 3.4, we can also interpret

$$7 - 5$$

as *how many more blocks* 7 blocks are than 5 blocks, or in other words, how many blocks you have to add to 5 blocks to get 7 blocks.

Notice that even the word *difference* that we use to describe the result of subtraction, $A - B$, contains the idea of comparison.

The *comparison* interpretation of subtraction is commonly used by shopkeepers when giving change after a purchase: the shopkeeper returns money to the patron, adding these returned amounts to the purchase price until he arrives at the amount of money that the patron gave him.

Exercises for Section 3.1

1. Use a number line to calculate $0 - (-2)$.

2. Use a number line to calculate $-3 - (-3)$.

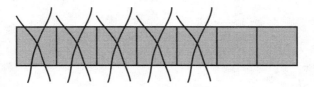

7 − 5 as 7 *take away* 5

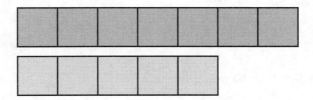

7 − 5 as *how many more* 7 is than 5

Figure 3.4: Two Views of Subtraction

Answers to Exercises for Section 3.1

1. See Figure 3.5.

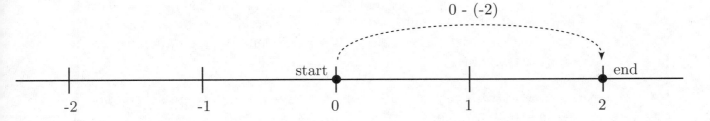

Figure 3.5: $0 - (-2) = 2$

2. See Figure 3.6

Problems for Section 3.1

1. Use a number line to calculate $3.8 + 1.9$.

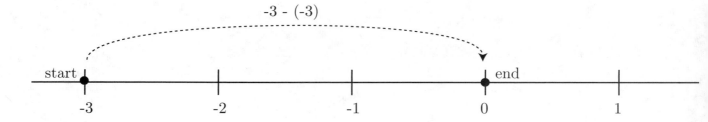

Figure 3.6: $-3 - (-3) = 0$

2. Use a number line to calculate $1.2 - 2.5$.

3. When a patron of a store gives the shopkeeper $\$A$ for a $\$B$ purchase, we can think of the change owed to the patron as what is left from $\$A$ when $\$B$ are taken away. But when the shopkeeper makes change, he often starts by naming $\$B$, and he then proceeds to hand the patron money, adding on the amounts until he reaches $\$A$.

   (a) Describe in detail how the shopkeeper will use the method above to give the patron change from a $20 bill on a $17.63 purchase.

   (b) In general, explain why a patron will always get the correct amount of change when the shopkeepers' method for making change is used. In particular, reconcile the method used by shopkeepers with the view that change is what is left from $\$A$ when $\$B$ are taken away.

4. Describe two situations other than change-making where subtraction is interpreted as comparison.

5. Make several copies of the number line in Figure 3.7, and use them to help you compare and contrast the following expressions:

$$-A + B$$
$$-(A + B)$$
$$-A - B$$
$$-(A - B)$$

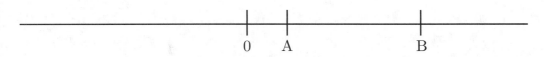

Figure 3.7: A Number Line

3.2 Why the Standard Algorithms for Adding and Subtracting Decimals Work

In school, we all learned the standard paper and pencil methods for adding, subtracting, multiplying, and dividing numbers. But have you ever wondered why they work? All these standard methods were invented by people, separately, in many different places throughout the world, and at many different times. How did the inventors know that their clever methods would yield correct results? Just as you can't bake a cake by randomly combining flour, sugar, butter, and eggs, you also can't come up with the correct answer to an addition problem by randomly combining the numbers involved. The inventors of the methods we use in arithmetic today had to think deeply to derive theses methods. We can now benefit from these methods, but as a teacher, you should not take the standard methods of arithmetic for granted. You should know why these methods work, and what the reasoning is that lies behind them. Because only when you understand that mathematics is based on sensible reasoning will you be able to teach mathematics so that it makes sense.

What is an Algorithm?

An **algorithm** is a method or a procedure for carrying out a calculation.

When you bake a cake, you probably follow a recipe. A cake recipe is a kind of algorithm: it is a step-by-step procedure for how to take various ingredients and turn them into a cake. In the same way, we have step-by-step procedures for adding numbers, subtracting numbers, multiplying numbers, and dividing numbers. Just as there is mystery in how flour, eggs, butter, and sugar can combine to make cake, there is mystery in how the standard algorithms come up with correct answers to arithmetic problems. Somehow, we throw numbers in, mix them up in a specific way, and out comes—cake? No, but out comes the correct answer to the addition problem. In this book, we won't examine the mysteries of cake-baking, but we will uncover the mysteries of the algorithms of arithmetic. In fact, they aren't mysteries at all, but very clever, efficient ways of calculating that make complete sense once you know how to look at them.

The Addition Algorithm

When we use the standard algorithm to add whole numbers or decimals, such as $34.5 + 7.89$ or $149 + 85$, we put the numbers one under the other, lining up the decimal points, and then we add column by column, *regrouping* as needed. When adding, we **regroup** when the digits in a column add to 10 or more, and we indicate regrouping by placing a small 1 at the top of the next column to the left. Regrouping is also called **trading** or **carrying**.

$$
\begin{array}{r}
\overset{1\;1}{34}.50 \\
+\ 7.89 \\
\hline
42.39
\end{array}
\qquad\qquad
\begin{array}{r}
\overset{1\;1}{149} \\
+\ 85 \\
\hline
234
\end{array}
$$

Notice that even though you don't see any decimal points in $149 + 85$, they are still lined up because we could also write this addition problem as $149.0 + 85.0$. Why do we line the decimal points up like that? Why do we add column by column? And why do we regroup? The answers to these questions lie in the interpretation of addition as *combining*, and in place value.

Working With Physical Objects To Understand the Addition Algorithm for Whole Numbers

Consider the sum $149 + 85$. We can represent this sum as the total number of toothpicks when 149 toothpicks are combined with 85 toothpicks. In order to analyze the addition algorithm, represent the decimals 149 and 85 with bundles as described in Section 2.3. Represent 149 toothpicks as 1 bundle of one hundred (which is ten bundles of ten), 4 bundles of ten, and 9 individual toothpicks, as shown in Figure 3.8. Similarly, represent 85 toothpicks as 8 bundles of ten and 5 individual toothpicks. When all these toothpicks are combined, how many are there? Adding like bundles, there are:

> 14 individual toothpicks (9 plus 5 more)

> 12 bundles of 10 toothpicks (4 bundles of ten plus 8 more bundles of ten)

and 1 bundle of 100 toothpicks.

Symbolically, we can write this as:

$$
\begin{array}{rrrrr}
1(100) & + & 4(10) & + & 9(1) \\
& + & 8(10) & + & 5(1) \\
\hline
= 1(100) & + & 12(10) & + & 14(1)
\end{array}
$$

So there are

$$1(100) + 12(10) + 14(1)$$

toothpicks, but this expression is not the expanded form of a number in ordinary decimal notation because there are 12 tens and 14 ones. This is where we regroup our bundles. *The regrouping of bundles of toothpicks is the physical representation of the regrouping process in the addition algorithm.*

To regroup the bundled toothpicks, convert the 14 individual toothpicks into 1 bundle of ten toothpicks and 4 individual toothpicks. This 1 bundle of ten is the small "carried" 1 we write above the 4 in the standard procedure. That small 1 really stands for 1 ten.

$$
\begin{array}{r}
^{1\,1} \\
14\ 9 \\
+85 \\
\hline
234
\end{array}
$$

Likewise, regroup the 12 bundles of 10 toothpicks into 1 bundle of 100 (namely 10 bundles of ten) and 2 bundles of ten. This 1 bundle of 100 is the small "carried" 1 we write above the 1 in the standard procedure. That small 1 really stands for 1 hundred. When you collect like bundles, you will have 2 bundles of a hundred, 3 bundles of ten, and 4 individual toothpicks. Here is the same example written symbolically:

$$
\begin{array}{rrrrr}
1(100) & + & 4(10) & + & 9(1) \\
& + & 8(10) & + & 5(1) \\
\hline
= 1(100) & + & \cancel{12(10)} & + & \cancel{14(1)} \\
& + & 1(10) & + & 4(1) \\
+1(100) & + & 2(10) & & \\
\hline
= 2(100) & + & 3(10) & + & 4(1)
\end{array}
$$

14 ones become 1 ten and 4 ones
12 tens become 1 hundred and 2 tens

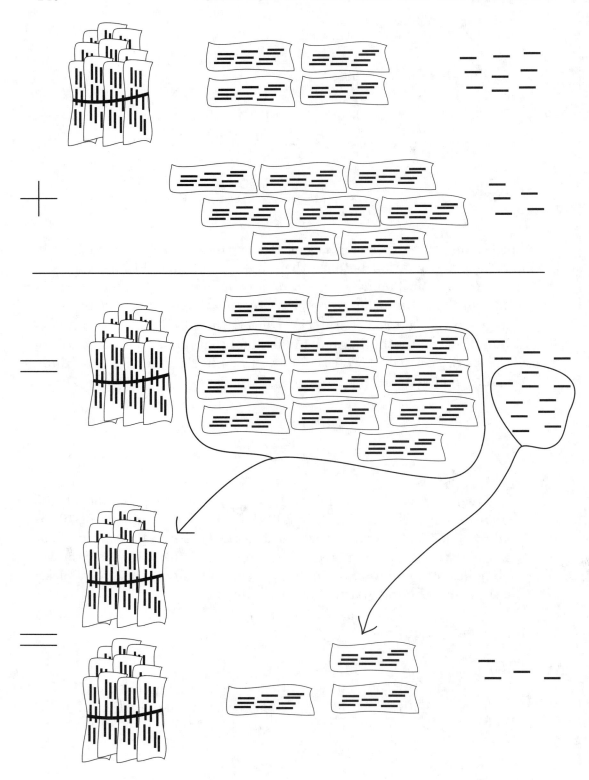

Figure 3.8: Regrouping $149 + 85$ toothpicks

We can also write this regrouping in equation form:

$$1(100) + 12(10) + 14(1) \quad = \qquad 1(100) \qquad + \quad 10(10) + 2(10) \quad + \quad 10(1) + 4(1)$$

$$= \quad 1(100) + 1(100) \quad + \quad 2(10) + 1(10) \quad + \qquad 4(1)$$
$$= \qquad 2(100) \qquad + \qquad 3(10) \qquad + \qquad 4(1)$$
$$= \qquad 234$$

(The diagonal arrows show you how the 10 tens become 1 hundred and the 10 ones become 1 ten.) Notice that these equations represent symbolically the *physical actions* of regrouping the toothpicks.

The key point is that the standard addition algorithm is just a way to condense the information in equations like the ones shown, and therefore is a way to quickly and efficiently record the physical action of adding and regrouping actual objects, such as toothpicks. This is why the standard addition algorithm gives us correct answers to addition problems.

The Addition Algorithm for Decimals

When we use the addition algorithm to add decimals, we use the same procedure as for whole numbers. The first step in the algorithm is to line up the decimal points. Although the decimal points are usually not shown in whole numbers, this step also occurs when adding whole numbers because the places where the decimal points would be are lined up. Why do we line up the decimal points? We can see why this makes sense by working with bundled toothpicks. The key lies in place value.

As we saw in Section 2.3, bundled toothpicks can be used to represent (finite) decimals, as long as the meaning of 1 toothpick is interpreted suitably. To represent the sum .834+6.7 with toothpicks, let 1 toothpick represent $\frac{1}{1000}$. Then a bundle of 10 toothpicks represents $\frac{1}{100}$, a bundle of 100 toothpicks (a bundle of 10 tens) represents $\frac{1}{10}$ and a bundle of 1000 toothpicks (a bundle of 10 hundreds) represents 1. So .834 is then represented by

> 8 bundles of 100 toothpicks,
>
> 3 bundles of 10 toothpicks and
>
> 4 individual toothpicks,

while 6.7 is represented by

6 bundles of 1000 toothpicks and

7 bundles of 100 toothpicks.

Notice that we chose 1 toothpick to represent $\frac{1}{1000}$ in *both* .834 and 6.7. In this way, when we use the bundled toothpicks to represent the sum .834+6.7, thousandths will be added to thousandths, hundredths to hundredths, tenths to tenths and ones to ones. If 1 toothpick were to represent $\frac{1}{1000}$ in .834 but $\frac{1}{10}$ in 6.7, then we would add thousandths to tenths and hundredths to ones, which wouldn't make any sense. This would be like treating a penny as $\frac{1}{100}$ of a dollar in one setting and $\frac{1}{10}$ of a dollar in another setting. *The consistent choice for the meaning of 1 toothpick when representing both decimals has the same effect as lining up the decimal points of the two decimals. Decimal points must be lined up so that like terms will be added*—tens to tens, ones to ones, tenths to tenths, hundredths to hudredths, and so on.

Once the decimal points have been lined up, the addition algorithm proceeds in the same way as if the decimal points weren't there, and the explanation for why the algorithm works is the same as for whole numbers. You can write 0s for the blank places if you like.

$$
\begin{array}{r}
{}^{1}.834 \\
+\ 6.7 \\
\hline
\end{array}
\qquad
\begin{array}{r}
{}^{1}.834 \\
0.834 \\
+\ 6.700 \\
\hline
\end{array}
$$

The Subtraction Algorithm

Now let's turn to subtraction. When we subtract decimals, such as $319-148.2$ or $142-83$, using the standard paper and pencil algorithm, we first put the numbers one under the other, lining up the decimal points. Then we subtract column by column, *regrouping* as needed so that we can subtract the numbers in a column without a negative number resulting. In subtraction, we **regroup** when the digit at the top of a column is less than the digit below it. We indicate regrouping by crossing out the digit at the top of the next column to the left, replacing this digit with the digit that is 1 less, and replacing the digit at the top of the original column with 10 plus this digit (for example, 2 is replaced with 12). If we can't carry this out because there is a 0 in the next column to the left, then we keep moving to the left, crossing out 0s until we come to a digit that is greater than 0. We cross this non-zero digit out and replace it with the digit that is 1 less, we replace all the intervening 0s with 9s, and, as before, we replace the digit at the top of the original column with 10 plus this digit. Regrouping is also called **trading** or **borrowing**.

$$
\begin{array}{r}
319.0 \\
-148.2 \\
\end{array}
\quad \rightarrow \quad
\begin{array}{r}
{\scriptstyle 2\ 11\ 8\ 10} \\
\cancel{3}\,\cancel{1}\,\cancel{9}.\,\cancel{0} \\
-1\,4\,8\,.2 \\
\hline
1\,7\,0\,.8 \\
\end{array}
$$

$$
\begin{array}{r}
142 \\
-83 \\
\end{array}
\quad \rightarrow \quad
\begin{array}{r}
{\scriptstyle 0\ 13\,12} \\
\cancel{1}\,\cancel{4}\,\cancel{2} \\
-\ 8\,3 \\
\hline
5\,9 \\
\end{array}
$$

$$
\begin{array}{r}
100.2 \\
-5.3 \\
\end{array}
\quad \rightarrow \quad
\begin{array}{r}
{\scriptstyle 0\ 9\ 9\ 12} \\
\cancel{1}\,\cancel{0}\,\cancel{0}.\,\cancel{2} \\
-\ 5\,.3 \\
\hline
9\,4\,.9 \\
\end{array}
$$

As with the addition algorithm, we want to understand why this procedure makes sense. We will explain why by interpreting subtraction as *taking away* and by considering place value.

Working With Physical Objects To Understand The Subtraction Algorithm

As with addition, the subtraction algorithm, and in particular, the regrouping process can be modeled with bundles of toothpicks. Consider the difference $142 - 83$. To represent this with bundled toothpicks, start with 142 toothpicks in 1 bundle of a hundred, 4 bundles of ten, and 2 individual toothpicks. How many toothpicks will be left when we take 83 toothpicks away? When we try to take 3 individual toothpicks away from the 142 toothpicks, we first need to do some unbundling. *This unbundling is a physical representation of the regrouping process.* Is is illustrated in Figure 3.9. One of the 4 bundles of ten can be unbundled and added to the individual toothpicks. This results in 1 bundle of a hundred, 3 bundles of ten and 12 individual toothpicks. In

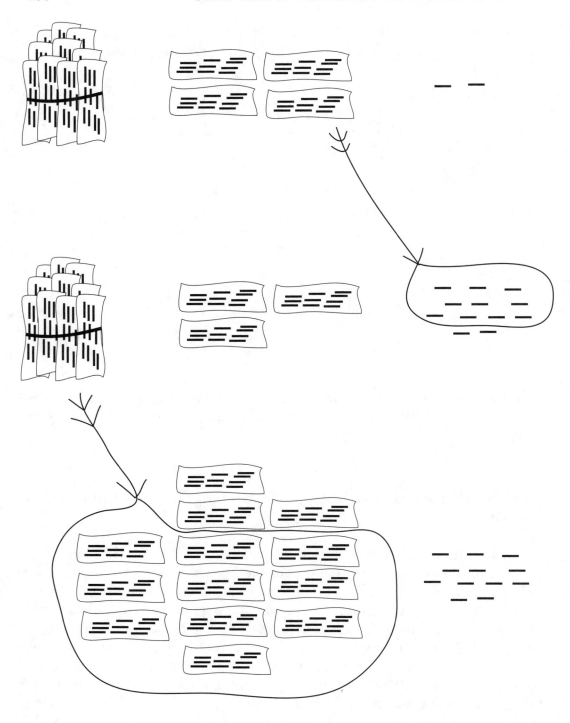

Figure 3.9: Regrouping 142 toothpicks

equations:

$$
\begin{aligned}
142 &= 1(100) &+& & 4(10) &&+& & 2(1) \\
&= 1(100) &+& 3(10) + 1(10) &&+& & 2(1) \\
&= 1(100) &+& & 3(10) &&+& 10(1) + 2(1) \\
&= 1(100) &+& & 3(10) &&+& & 12(1)
\end{aligned}
$$

Now there won't be any problem taking 3 individual toothpicks away, but what about taking away 80, namely 8 bundles of ten? There are only 3 bundles of ten. So we must unbundle the 1 bundle of a hundred as 10 bundles of 10 and combine these 10 bundles of ten with the 3 bundles of ten to make 13 bundles of ten. In equations, continuing from above:

$$
\begin{aligned}
142 &= 1(100) &+& & 3(10) && + & 12(1) \\
&= && & 10(10) + 3(10) && + & 12(1) \\
&= && & 13(10) & 12(1)
\end{aligned}
$$

Now there's no difficulty taking 3 individual toothpicks away from the 12 individual toothpicks and 8 bundles of ten away from the 13 bundles of ten. Notice that the physical actions of moving the toothpicks correspond exactly to the regrouping process that takes place in the standard subtraction algorithm. In expanded form, the subtraction problem $142 - 83$ can now be rewritten as

$$
\begin{array}{r}
13(10) \;+\; 12(1) \\
-[8(10) \;+\; 3(1) \\
\hline
= 5(10) \;+\; 9(1) \quad = 59
\end{array}
$$

The key point is that the standard subtraction algorithm is just a way to condense the information in equations like the ones above, and therefore it is a quick and efficient way to record the physical actions of regrouping and then taking away actual objects, such as toothpicks. This is why the standard subtraction algorithm gives us correct answers to subtraction problems.

Understanding The Subtraction Algorithm For Decimals

As with the addition algorithm for decimals, we can generally still represent decimal subtraction problems with bundles of toothpicks by an appropriate interpretation of 1 toothpick. As before, decimal points are lined up so that like terms are subtracted—tens from tens, ones from ones, tenths from tenths, hundredths from hudredths, and so on.

Class Activity 3A: Understanding the Standard Algorithms for Adding and Subtracting

Class Activity 3B: Regrouping in Other Situations

Class Activity 3C: Subtracting Across Zeros

Exercises for Section 3.2 on the Addition and Subtraction Algorithms

1. Review the meanings of the following terms:

   algorithm

   sum

   difference

   regroup (in addition and in subtraction)

2. Explain why the standard algorithm for addition gives the correct answer to $1.8 + .67$. Illustrate the procedure with toothpicks (or some other small object) and explain the regrouping process. Write corresponding equations with the numbers in expanded form.

3. Explain why the standard algorithm for subtraction gives the correct answer to $104 - 69$. Illustrate the procedure with bundled toothpicks (or some other small object) and explain the regrouping process. Write corresponding equations with the numbers in expanded form.

4. Why do we line up decimal points before adding or subtracting decimals?

5. Ellie solves the subtraction problem $2.5 - .13$ with toothpicks. She represents 2.5 with 2 bundles of 10 toothpicks and 5 individual toothpicks, and she represents $.13$ with 1 bundle of 10 toothpicks and 3 individual toothpicks. Ellie gets the answer 1.2. Is she right?

6. A store buys action figures in boxes. Each box contains 50 bags, and each bag contains 6 action figures. At the beginning of the month, the store has

   7 unopened boxes, 15 unopened bags, and 3 individual action figures.

At the end of the month, the store has

> 2 unopened boxes, 37 unopened bags, and 5 individual action figures.

How many action figures did the store sell during the month (assuming they got no additional shipments of action figures)? Write your answer in terms of boxes, bags, and individual action figures. *Work with boxes, bags, and individuals in a sort of expanded form and use regrouping* to solve this problem.

Answers to Exercises for Section 3.2 on the Addition and Subtraction Algorithms

1. See text.

2. Represent 1.8 by 1 bundle of 100 toothpicks, 8 bundles of 10 toothpicks and 0 individual toothpicks. Similarly, represent .67 by 6 bundles of 10 toothpicks and 7 individual toothpicks. Working symbolically with expanded forms, the corresponding equations are:

$$
\begin{array}{rcccc}
 & 1(1) & + & 8(\frac{1}{10}) & + & 0(\frac{1}{100}) \\
+ & & & 6(\frac{1}{10}) & + & 7(\frac{1}{100}) \\
\hline
= & 1(1) & + & 14(\frac{1}{10}) & + & 7(\frac{1}{100})
\end{array}
$$

and

$$
\begin{array}{rcccccc}
1(1) + 14(\frac{1}{10}) + 7(\frac{1}{100}) & = & 1(1) & + & 10(\frac{1}{10}) + 4(\frac{1}{10}) & + & 7(\frac{1}{100}) \\
& = & 1(1) + 1(1) & + & 4(\frac{1}{10}) & + & 7(\frac{1}{100}) \\
& = & 2(1) & + & 4(\frac{1}{10}) & + & 7(\frac{1}{100})
\end{array}
$$

3. Here's the regrouping process, shown in gory detail with equations in expanded form:

$$
\begin{array}{rcccccc}
104 & = & 1(100) & + & 0(10) & + & 4(1) \\
& = & & & 10(10) + 0(10) & + & 4(1) \\
& = & & & 10(10) & + & 4(1) \\
& = & & & 9(10) + 1(10) & + & 4(1) \\
& = & & & 9(10) & + & 10(1) + 4(1) \\
& = & & & 9(10) & + & 14(1)
\end{array}
$$

4. See text.

5. No, Ellie's answer is not correct. 1 toothpick must represent the same amount when representing both 2.5 and .13. Ellie should think of 1 toothpick as representing $\frac{1}{100}$ in both cases. Then 2.5 is represented by 2 bundles of 100 toothpicks and 5 bundles of 10 toothpicks, while .13 is represented by 1 bundle of 10 toothpicks and 3 individual toothpicks. Now Ellie should be able to see that she'll need to regroup in order to subtract. It might help Ellie to think in terms of money: 2.5 and .13 can be represented by \$2.50 and \$.13. Ellie's way of using the toothpicks would be like saying that a dime is equal to a penny.

6. We must solve:

$$
\begin{array}{r}
7 \text{ boxes } + 15 \text{ bags } + 3 \text{ individual} \\
-(2 \text{ boxes } + 37 \text{ bags } + 5 \text{ individual }) \\
\hline
\end{array}
$$

We can solve this by first regrouping the 7 boxes, 15 bags, and 3 individual action figures. If we open one of the bags, then there is one less bag, but 6 more individual figures, so there are

    7 boxes, 14 bags, and 9 individual action figures.

If we open one of the boxes, then there is one less box, but 50 more bags of action figures, so there are

    6 boxes, 64 bags, and 9 individual action figures.

It's still the same number of action figures, they are just arranged in a different way. In equation form we can write this as:

$$
\begin{aligned}
7 \text{ boxes } + 15 \text{ bags } + 3 \text{ individual} \quad &= \quad 7 \text{ boxes } + 14 \text{ bags } + (6+3) \text{ individual} \\
&= \quad 6 \text{ boxes } + (50+14) \text{ bags } + 9 \text{ individual} \\
&= \quad 6 \text{ boxes } + 64 \text{ bags } + 9 \text{ individual}
\end{aligned}
$$

Now we are ready to subtract the 2 boxes, 37 bags, and 5 individual action figures:

$$
\begin{array}{r}
6 \text{ boxes } + 64 \text{ bags } + 9 \text{ individual} \\
-(2 \text{ boxes } + 37 \text{ bags } + 5 \text{ individual }) \\
\hline
4 \text{ boxes } + 27 \text{ bags } + 4 \text{ individual}
\end{array}
$$

So a total of 4 boxes, 27 bags, and 4 individual action figures were sold during the month.

Problems for Section 3.2 on the Addition and Subtraction Algorithms

1. In order to add $153 + 87$, Josh wants to line the numbers up this way:

$$153$$
$$+\ \ 87$$

   Explain to Josh why his method doesn't work, and explain why the correct way of lining up the numbers makes sense.

2. Allie solves the subtraction problem $304 - 9$ as follows:

$$\overset{2}{\cancel{3}}\ 0\ \overset{14}{\cancel{4}}$$
$$-9$$
$$\overline{2\ 0\ 5}$$

   Explain to Allie what is wrong with her method, and explain why the correct method makes sense.

3. On a space shuttle mission, a certain experiment is started 2 days, 14 hours and 30 minutes into the mission. The experiment takes 1 day, 21 hours and 47 minutes to run. When will the experiment be completed? Give your answer in days, hours and minutes into the mission. Do your work in a sort of expanded form, in other words, work with

$$2(\text{days}) + 14(\text{hours}) + 30(\text{minutes})$$

   and

$$1(\text{day}) + 21(\text{hours}) + 47(\text{minutes}),$$

   and *regroup between days, hours, and minutes* to solve this problem.

4. We can write dates and times in a sort of expanded form. For example, October 4th, 6:53pm can be written as

$$4(\text{days}) + 18(\text{hours}) + 53(\text{minutes})$$

   (in some circumstances you might want to include the month and the year too). How long is it from 3:27pm on October 4, to 7:13am on October 19? Give your answer in days, hours, minutes. Solve this by working in expanded form and *regroup between days, hours, and minutes* to solve this problem.

5. Erin wants to figure out how much time it is from 9:45 am to 11:30 am. Erin does the following:

$$
\begin{array}{r}
1\,\overset{0}{\cancel{1}}\!:\!\overset{12}{\cancel{3}}\overset{1}{0} \\
-9:45 \\
\hline
1:85
\end{array}
$$

and gives 1 hour and 85 minutes as the answer. Is Erin right? If not, explain what is wrong with her method and show how to *modify her method* to make it correct. In addition, solve the problem in another way, explaining your method.

6. Here's how Mo solved the subtraction problem $635 - 813$:

$$
\begin{array}{r}
635 \\
-813 \\
\hline
-222
\end{array}
$$

Mo did this by working from right to left, saying:

$$
\begin{aligned}
5 - 3 &= 2 \\
3 - 1 &= 2 \\
6 - 8 &= -2.
\end{aligned}
$$

(a) Is Mo's answer right?

(b) Solve

$$
\begin{array}{r}
6(100) \ + \ 3(10) \ + \ 5(1) \\
- \ [8(100) \ + \ 1(10) \ + \ 3(1)] \\
\hline
\end{array}
$$

by working with expanded forms. Discuss how Mo's work compares to your work with expanded forms.

Is Mo's answer right? If not, why not? Use expanded forms to analyze Mo's work. If Mo's answer is not correct, find a way to *modify* Mo's work to make it correct (do not just tell Mo a different correct way of doing the problem, use what Mo has already done and modify it). Explain your answers carefully.

7. The subtraction algorithm described in the text is not the only standard subtraction algorithm. Some people use the following algorithm

instead: Line up the numbers as in the standard algorithm, and subtract column by column, proceeding from right to left. The only difference between this new algorithm and our standard one is in regrouping. To regroup with the new algorithm, move one column to the left and cross out the digit at the *bottom* of the column, replacing it with that digit *plus 1* (so replace an 8 with a 9, replace a 9 with a 10, etc.). Then, as in the standard algorithm, replace the digit at the top of the original column with that number plus 10. The following example shows the steps of this new algorithm.

$$
\begin{array}{r}
132 \\
-79 \\
\hline
\end{array}
\quad \rightarrow \quad
\begin{array}{r}
13\overset{12}{\not2} \\
-\,\overset{8}{\not7}9 \\
\hline
3
\end{array}
\quad \rightarrow \quad
\begin{array}{r}
1\overset{13}{\not3}\overset{12}{\not2} \\
-\,\overset{1}{\not0}\overset{8}{\not7}9 \\
\hline
53
\end{array}
$$

(a) Use the new algorithm to solve $524 - 198$ and $1003 - 95$. Verify that the new algorithm gives correct answers.

(b) Explain why it makes sense that this new algorithm gives correct answers to subtraction problems. What is the reasoning behind this new algorithm?

(c) What are some advantages and disadvantages of this new algorithm compared to our standard one?

8. The subtraction algorithm described in the text is not the only standard subtraction algorithm. Here is another subtraction algorithm, called *adding the complement*.   For a 3-digit whole number, N, the **complement** of N is $999 - N$. For example, the complement of 486 is

$$999 - 486 = 513.$$

Notice that regrouping is never needed to calculate the complement of a number. To use the *adding the complement* algorithm to subtract a 3-digit whole number, N, from another 3 digit whole number, you start by adding the complement of N rather than subtracting N. For example, to solve

$$723 - 486,$$

first add the complement of 486:

$$
\begin{array}{r}
723 \\
+513 \\
\hline
1236
\end{array}
$$

Then cross out the 1 in the thousands column, and add 1 to the resulting number:

$$\cancel{1}236 \quad \rightarrow \quad 236 + 1 = 237.$$

Therefore, according to the *adding the complement* algorithm, $723 - 486 = 237$.

(a) Use the *adding the complement* algorithm to calculate $301 - 189$ and $295 - 178$. Verify that you get the correct answer.

(b) Explain why the *adding the complement* algorithm gives you the correct answer to any three digit subtraction problem. In order to explain this, focus on how the original problem and the addition problem in *adding the complement* are related. For example, how are the problems $723 - 486$ and $723 + 513$ related? Work with the *complement* relationship, $513 = 999 - 486$, and notice that $999 = 1000 - 1$.

(c) What are some advantages and disadvantages of the *adding the complement* algorithm compared to our standard algorithm?

3.3   Adding And Subtracting Fractions

In the last section, we learned why the standard addition and subtraction algorithms for whole numbers and decimals make sense. There are also standard algorithms for adding and subtracting fractions, and in this section, we will study why these make sense. Because they are defined in terms of addition, we will also study mixed numbers (numbers such as $2\frac{3}{4}$) and the algorithm for converting a mixed number to an improper fraction.

If two fractions have the same denominator, then you can add or subtract these fractions by adding or subtracting the numerators and leaving the denominator unchanged. For example,

$$\frac{4}{15} + \frac{7}{15} = \frac{4+7}{15} = \frac{11}{15}.$$

Why does this make sense? We can represent the sum

$$\frac{4}{15} + \frac{7}{15}$$

as the total amount of pie you will have if you start with $\frac{4}{15}$ of a pie and get $\frac{7}{15}$ more of the pie. But $\frac{4}{15}$ of a pie is represented by 4 pieces when the pie is divided into 15 equal pieces, and $\frac{7}{15}$ is represented by 7 of those pieces. So if you have 4 pieces of pie and you get 7 more, then you will have $4 + 7 = 11$ pieces all together, as you see in Figure 3.10. What kind of pieces are they? All 11 pieces are fifteenths of a pie. Therefore you have $\frac{11}{15}$ of a pie in all.

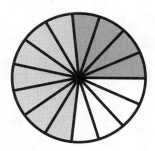

Figure 3.10: $\frac{4}{15} + \frac{7}{15} = \frac{11}{15}$

How do we add or subtract fractions that have different denominators? As with comparing fractions, first write your fractions with a common denominator, then add the numerators. *Any* common denominator will do, it does not have to be the least one. You can always produce a common denominator by multiplying the two denominators. So, to add

$$\frac{5}{6} + \frac{3}{8},$$

we can use the common denominator $6 \cdot 8$, which is 48, or, if we want to work with smaller numbers, we can use 24, which is a common denominator because $24 = 6 \cdot 4$ and $24 = 8 \cdot 3$.

$$\frac{5}{6} + \frac{3}{8} = \frac{40}{48} + \frac{18}{48} = \frac{58}{48}$$

or

$$\frac{5}{6} + \frac{3}{8} = \frac{20}{24} + \frac{9}{24} = \frac{29}{24}.$$

Since

$$\frac{58}{48} = \frac{29 \cdot 2}{24 \cdot 2} = \frac{29}{24},$$

the two ways of adding $\frac{5}{6} + \frac{3}{8}$ produce equal results. Figure 3.11 shows the addition of $\frac{5}{6}$ and $\frac{3}{8}$ with the common denominator 24, using a number line.

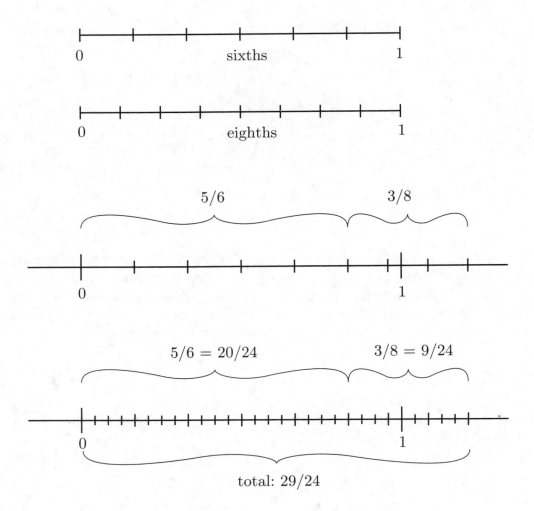

Figure 3.11: Using A Common Denominator To Add

In general, suppose

$$\frac{A}{B} \quad \text{and} \quad \frac{C}{D}$$

are two fractions (where $B \neq 0$ and $D \neq 0$). Then

$$\frac{A}{B} = \frac{A \cdot D}{B \cdot D}$$

and

$$\frac{C}{D} = \frac{C \cdot B}{D \cdot B},$$

and the latter two fractions

$$\frac{A \cdot D}{B \cdot D} \quad \text{and} \quad \frac{C \cdot B}{D \cdot B}$$

both have denominator equal to $B \cdot D$. Because both fractions are expressed in terms of like parts, we can add or subtract these fractions by adding or subtracting the numerators. Therefore in general:

$$\frac{A}{B} + \frac{C}{D} = \frac{A \cdot D + C \cdot B}{B \cdot D}$$

and

$$\frac{A}{B} - \frac{C}{D} = \frac{A \cdot D - C \cdot B}{B \cdot D}.$$

Mixed Numbers and Improper Fractions

A **mixed number** is a number that is written in the form

$$A\frac{B}{C},$$

where A, B, and C are whole numbers, and $\frac{B}{C}$ is a proper fraction (i.e., the numerator is less than the denominator). So

$$2\frac{3}{4} \quad \text{and} \quad 5\frac{7}{8}$$

are mixed numbers.

The mixed number

$$A\frac{B}{C}$$

just stands for

$$A + \frac{B}{C}.$$

Using this definition of mixed numbers, and the fact that a whole number A is equal to the fraction $\frac{A}{1}$, you can show that every mixed number can be written as an improper fraction as follows:

$$A\frac{B}{C} = \frac{A \cdot C + B}{C}.$$

So

$$2\frac{3}{4} = \frac{2 \cdot 4 + 3}{4} = \frac{11}{4}.$$

When is Combining Not Adding?

You've probably heard the old admonishment: "you can't add apples and oranges." This is especially important to remember when adding and subtracting fractions, because sometimes fractions of quantities that are to be combined refer to *different wholes*.

We can represent the sum

$$3 + 4$$

as the total number of apples when 3 apples are combined with 4 apples. But if you represent $3+4$ by combining 3 bananas and 4 blocks, as in Figure 3.12, then the best you can say is that you have 7 *objects* in all.

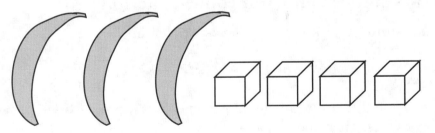

Figure 3.12: 3 Bananas and 4 Blocks

Similarly, we can represent the sum

$$\frac{1}{3} + \frac{1}{4}$$

as the fraction of a pie we will have in all if we have $\frac{1}{3}$ of the pie and then get another $\frac{1}{4}$ *of the same pie or an equivalent pie.* The $\frac{1}{3}$, the $\frac{1}{4}$ and the sum $\frac{1}{3} + \frac{1}{4} = \frac{7}{12}$ *all refer to the same pie,* or to copies of an equivalent pie.

But suppose we have a small pie and a large pie, and that the small pie is divided into 3 pieces and the large pie is divided into 4 pieces, as shown in Figure 3.13. If we get one piece of each pie, does this represent the sum

$$\frac{1}{3} + \frac{1}{4}?$$

No, it does not, because the $\frac{1}{3}$ and the $\frac{1}{4}$ refer to *different wholes* that are not equivalent. All we can say is that we have 2 pieces of pie. But we can't say that we have $\frac{7}{12}$ of a pie—because what pie would the $\frac{7}{12}$ refer to?

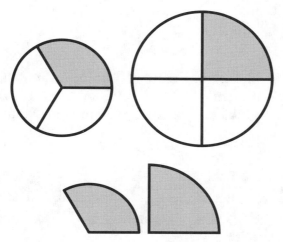

does the combined amount represent

$$\frac{1}{3} + \frac{1}{4}?$$

Figure 3.13: $\frac{1}{3}$ of a Little Pie and $\frac{1}{4}$ of a Big Pie

Class Activity 3D: Fraction Addition and Subtraction

Class Activity 3E: Mixed Numbers

Exercises for Section 3.3 on Adding And Subtracting Fractions

1. We usually call the number .43 "forty-three hundredths" and not "four tenths and three hundredths". Why do these two mean the same thing?

2. Explain why it makes sense to add fractions according to the rule:

$$\frac{A}{B} + \frac{C}{D} = \frac{A \cdot D + C \cdot B}{B \cdot D}.$$

3. Jessica says that

$$\frac{1}{2} + \frac{2}{3} = \frac{3}{5}$$

and shows the picture in Figure 3.14 to prove it. What is the problem with Jessica's reasoning? Why is it not correct? Don't just explain how to do the problem correctly, explain what is faulty with Jessica's reasoning.

Figure 3.14: $\frac{1}{2} + \frac{2}{3} = \frac{3}{5}$?

4. The usual procedure for turning a mixed number, such as $7\frac{1}{3}$, into an improper fraction is this:

$$7\frac{1}{3} = \frac{7 \cdot 3 + 1}{3} = \frac{22}{3}.$$

Explain why this procedure makes sense. What is the logic behind this procedure?

Answers to Exercises for Section 3.3 on Adding And Subtracting Fractions

1. Notice that four tenths is the same as forty hundredths, so four tenths and three hundredths is forty hundredths and three hundredths, or forty three hundredths. With equations:

$$
\begin{aligned}
.43 &= \frac{4}{10} + \frac{3}{100} \\
&= \frac{40}{100} + \frac{3}{100} \\
&= \frac{43}{100}.
\end{aligned}
$$

2. See text.

3. Jessica's pictures show that $\frac{1}{2}$ of the two-block bar combined with $\frac{2}{3}$ of the three-block bar does indeed make up $\frac{3}{5}$ of a five-block bar. The problem is that Jessica is using three different *wholes*. When we add fractions, such as $\frac{1}{2} + \frac{2}{3}$, both fractions in the sum, and the sum itself, should all refer to the same whole.

4. Remember that $7\frac{1}{3}$ stands for $7 + \frac{1}{3}$, and that $7 = \frac{7}{1}$. To calculate

$$
7\frac{1}{3} = \frac{7}{1} + \frac{1}{3}
$$

as a fraction, we need a common denominator:

$$
\begin{aligned}
7\frac{1}{3} &= 7 + \frac{1}{3} \\
&= \frac{7}{1} + \frac{1}{3} \\
&= \frac{7 \cdot 3}{1 \cdot 3} + \frac{1}{3} \\
&= \frac{7 \cdot 3 + 1}{3} \\
&= \frac{22}{3}.
\end{aligned}
$$

Notice that the next-to-last step shows the procedure for turning a mixed number into an improper fraction. Therefore this procedure is really just a shorthand way to give 7 and $\frac{1}{3}$ a common denominator and add them.

Problems for Section 3.3 on Adding And Subtracting Fractions

1. John says $\frac{2}{3} + \frac{2}{3} = \frac{4}{6}$ and uses the picture in Figure 3.15 as evidence. Discuss what is wrong with John's reasoning. What underlying misconception does John have? *Don't* just explain how to do the problem correctly, explain what is faulty with John's reasoning.

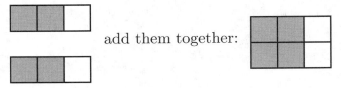

add them together:

Figure 3.15: $\frac{2}{3} + \frac{2}{3} = \frac{4}{6}$

2. Suppose you start with a fraction and you add 1 to the numerator and to the denominator. For example, if you started with $\frac{2}{3}$, then you'd get a new fraction $\frac{2+1}{3+1} = \frac{3}{4}$. Is this procedure of adding 1 to the numerator and the denominator the same as adding the number 1 to the original fraction? (For example, is $\frac{2+1}{3+1}$ equal to $\frac{2}{3} + 1$?) Explain!

3. In the first part of the season, the Bluejays play 18 games and win 7, while the Robins play 3 games and win 1. In the second part of the season, the Bluejays play 2 games and win 1, while the Robins play 17 games and win 8.

   (a) Who won a larger fraction of their games in the first part of the season?

   (b) Who won a larger fraction of their games in the second part of the season?

   (c) Who won a larger fraction of their games overall?

   (d) Compare your answers in parts (a), (b) and (c). You may be surprised at the comparison. (If you are not surprised, then you should probably rethink your work in part (c).)

   (e) Is the fraction of games won overall by the Bluejays equal to the sum of the fractions of games won by the Bluejays in the first and second parts of the season?

(f) What are the different *wholes* that are associated to the fractions in this problem?

(g) Write a paragraph on the relationship between your answers to (e) and (f) and the discussion in the text on *combining* versus *adding*.

4. Denise says that $\frac{2}{3} - \frac{1}{2} = \frac{1}{3}$ and gives the reasoning indicated in Figure 3.16 to support her answer. Is Denise right? If not, what is wrong with her reasoning and how could you help her understand her mistake and fix it? *Don't just* explain how to solve the problem correctly, explain where Denise's reasoning is flawed and discuss what might be causing her misconception.

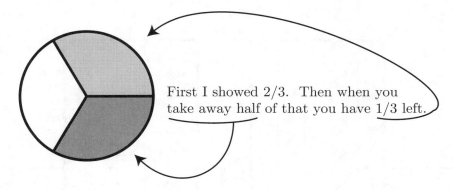

First I showed 2/3. Then when you take away half of that you have 1/3 left.

Figure 3.16: Denise's Idea For $\frac{2}{3} - \frac{1}{2}$

3.4 When Do We Add Percentages?

Every percentage can be expressed as a decimal or a fraction. So our discussions on addition of decimals and fractions covers the addition of percentages as well. In this brief section we will consider percent problems in which *combining* takes place but addition of percentages is nevertheless not appropriate.

How can there be a percent problem in which combining take place but addition of the percentages is not appropriate? This is the very same issue that occurs for fractions, and that was discussed in the last section, beginning on page 162. As with fractions, when you encounter a percent problem that involves combining, be aware of the *wholes* that the percentages refer to. *Addition of percentages is not appropriate when the wholes involved are not the same.*

Class Activity 3F: Should We Add These Percentages?

Exercises for Section 3.4 on Adding Percentages

1. Each of the pictures in Figure 3.17 shows two adjacent lots of land, lot A and lot B. In each case, 20% of lot A is shown shaded and 40% of lot B is shown shaded. What percent of the *combined amount* of lot A and lot B is shaded in each case?

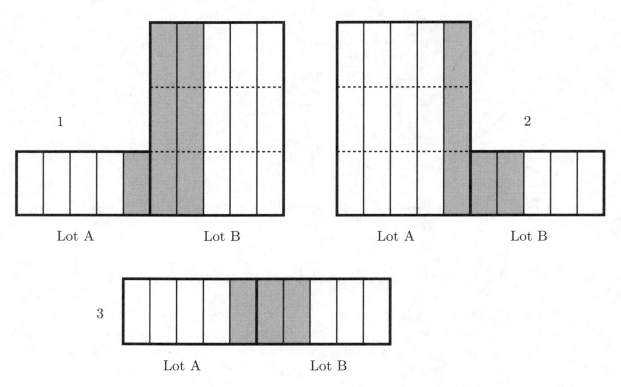

Figure 3.17: What Percent Of The Combined Amount Of Lots A And B Is Shaded?

Answers to Exercises for Section 3.4 on Adding Percentages

1. See Figure 3.18.

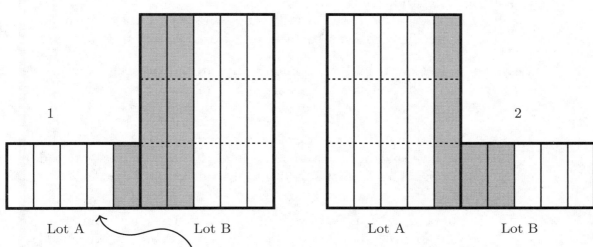

There are 20 small pieces of this size making up the combined lots A and B. Therefore each such piece is 1/20, which is 5%. Since 7 are shaded, that means 35% of the combined amount is shaded.

This time 5 pieces are shaded, each of which is 5% of the total. Therefore 25% of the total is shaded.

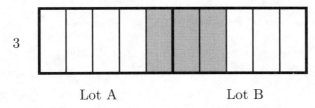

$3/10 = 30/100 = 30\%$ of the combined amount is shaded.

Figure 3.18: Various Percents Of The Combined Amount Of Lots A And B Are Shaded

Problems for Section 3.4 on Adding Percentages

1. Broad Street divides Popperville into an east side and a west side. 20% of the children on the east side of Popperville qualify for a reduced-price lunch. 30% of the children on the west side of Popperville qualify for a reduced-price lunch. Is it correct to calculate the percentage of children in all of Popperville who qualify for reduced-price lunch by adding 20% and 30% to get 50%? If the answer is no, why not? Explain in detail and calculate the correct percentage in at least two examples.

2. Anklescratch County and Kneebend County are two adjacent counties. In Anklescratch County, 30% of the miles of road have bike lanes, whereas in Kneebend County, only 10% of the miles of road have bike lanes.

   (a) Elliot says that in the two county area, 40% of the miles of road have bike lanes. Elliot found 40% by adding 30% and 10%. Is Elliot correct or not? Explain.

   (b) What percent of miles of road in the two county area have bike lanes if Anklescratch County has 200 miles of road and Kneebend County has 800 miles of road? What if it's the other way around and Anklescratch County has 800 miles of road whereas Kneebend County has 200 miles of road?

   (c) Ming says that 20% of the miles of road in the two county area have bike lanes because 20% is the average of 10% and 30%. Explain why Ming's answer could be either correct or incorrect. Under what circumstances will Ming's answer be correct?

3. There are two elementary schools in the town of South Elbow. In the first school, 20% of the children are Hispanic, in the second school 30% of the children are Hispanic. Write a brief essay about what you can and cannot tell from this data alone. Include examples to support your points.

3.5 Percent Increase and Percent Decrease

When a quantity increases or decreases, determining the amount of change is a simple matter of subtraction. However, in many situations, the actual value of the increase or decrease is less informative than the *percent* that this increase or decrease represents. For example, suppose that this year, there are 50 more children at Timothy Elementary School than there were last year. If Timothy Elementary only had 100 children last year, then 50 additional children is a huge increase. On the other hand, if Timothy Elementary had 500 children less year, an increase of 50 children is less significant. In this section we will study increases and decreases in quantities as *percents* rather than as fixed values.

If the value of a quantity goes up, then the increase in the quantity, figured as a percent of the original, is the **percent increase** of quantity. If a piece of furniture cost \$279 last week, and this week the same piece of furniture costs \$319, then the price went up by \$40. The percent increase of the cost of the furniture is the percent that this \$40 increase represents of the original cost, \$279. What percent of \$279 is \$40? Let $P\%$ be this percentage. Then

$$P\% \cdot 279 = 40,$$

so

$$P\% = \frac{40}{279} = .143 = 14.3\%.$$

Therefore \$40 is about 14% of \$279, so the price of the furniture increased by about 14%.

When a quantity decreases in value, there is the notion of a **percent decrease**. It is the decrease, figured as a percent of the original. If Lower Heeltoe got 215 inches of rainfall in 2000 and only 193 inches of rainfall in 2001, then Lower Heeltoe got 22 inches less rain in 2001 than in 2000. What percent of 215 inches is 22 inches? Let $P\%$ be this percentage. Then

$$P\% \cdot 215 = 22,$$

so

$$P\% = \frac{22}{215} = .102 = 10.2\%.$$

Therefore 22 inches is 10.2% of 215 inches. So the amount of rainfall in Lower Heeltoe decreased by 10.2% from 2000 to 2001.

Two Methods for Calculating Percent Increase or Decrease

The First Method

Our first method for calculating a percent increase or percent decrease uses the meaning of these terms. This was the method used in the examples above. In general, this method works as follows. Suppose a quantity changes from an amount A—the reference amount—to an amount B.

$$A \qquad \rightarrow \qquad B$$
$$\text{reference amount} \qquad \text{changed amount}$$

To calculate the percent increase or decrease in the quantity:

1. Calculate the *change*, C, in the quantity, namely either $B - A$ or $A - B$, whichever is positive (or 0).

2. Calculate the percent that the change C is of the reference amount A. If we call this percent $P\%$, then

$$P\% \cdot A = C,$$

so

$$P\% = \frac{C}{A}.$$

When you divide C by A you will get this percentage as a decimal. To write it as a percent, multiply by 100:

$$100 \times \frac{C}{A}.$$

The result is the percent increase or decrease of the quantity from A to B.

The Second Method

Our second method for calculating percent increase or decrease is usually more efficient. To derive this method, suppose that a quantity increases from an amount A to a larger amount B. If you calculate B as a percent of A, you will find that it is more than 100%. This makes sense because B is more than A, so B must represent more than 100% of A. *The amount that*

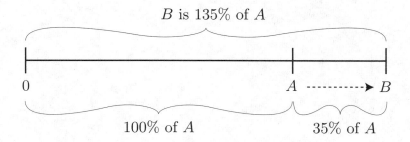

Figure 3.19: Calculating Percent Increase

B, calculated as a percent of A, is over 100%, is the percent increase from A to B. A picture, such as Figure 3.19, shows best why this is so.

The discussion above indicates why you can calculate the percent increase of a quantity from an amount A to a larger amount B this way:

1. Calculate B as a percent of A. You can do this by dividing B by A and multiplying the result by 100:

$$100 \times \frac{B}{A}.$$

2. Subtract 100%. The result is the percent increase of the quantity from A to B.

For example, suppose that the population of a city increases from 35,000 people to 44,000 people. What is the percent increase in the population?

$$44,000 \div 35,000 = 1.257 = 125.7\%.$$

Subtracting 100%, we determine that the population increased by 25.7%.

Similarly, if a quantity decreases from an amount A to a smaller amount B, then if you calculate B as a percent of A, you will find that it is less than 100%. This makes sense because B is less than A, so B must represent less than 100% of A. *The amount that B, calculated as a percent of A, is under 100%, is the percent decrease from A to B.* A picture, such as Figure 3.20, shows best why this is so.

The discussion above indicates why you can calculate the percent decrease of a quantity from an amount A to a smaller amount B this way:

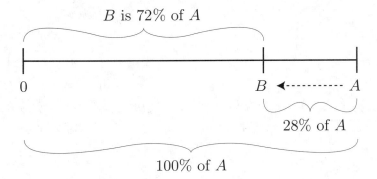

Figure 3.20: Calculating Percent Decrease

1. Calculate B as a percent of A. You can do this by dividing B by A and multiplying the result by 100:

$$100 \times \frac{B}{A}.$$

2. Subtract this percent from 100%. The result is the percent decrease of the quantity from A to B.

For example, suppose that the price of a computer dropped from \$899 to \$825. What is the percent decrease in the price of the computer?

$$825 \div 899 = .918 = 91.8\%.$$

$$100\% - 91.8\% = 8.2\%,$$

so the price of the computer decreased by about 8%.

Two Methods for Calculating Amounts when the Percent Increase or Decrease is Given

The First Method

When an amount and its percent increase are given, you can calculate the new amount by calculating the increase and adding this increase to the original amount. Similarly, when an amount and its percent decrease are given, you can calculate the new amount by calculating the decrease and subtracting this decrease from the original amount.

If gas costs $1.87 per gallon and if gas prices go up by 6%, then what will the new price of gas be? 6% of $1.87 is

$$.06 \times \$1.87 = \$.11,$$

so the new price of gas will be

$$\$1.87 + \$.11 = \$1.98$$

per gallon. On the other hand, if gas costs $1.87 per gallon and gas prices go down by 6%, then the new price of gas will be

$$\$1.87 - \$.11 = \$1.76.$$

The Second Method

Our second method for calculating amounts when percent increases or decreases are given uses the same idea as the second method above for calculating percent increases and decreases. A benefit of this method is that it can also be used to calculate the *initial amount* when the percent increase or decrease and the *final amount* are known.

As seen in Figure 3.21, if an amount A increases by 35% to an amount

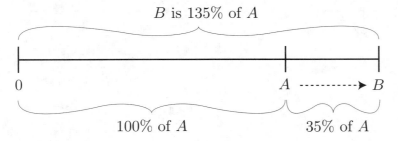

Figure 3.21: Calculating Amounts from the Percent Increase

B, then

$$B \text{ is } 135\% \text{ of } A.$$

In general, if an amount A inceses by $P\%$ to an amount B, then

$$B \text{ is } (100 + P)\% \text{ of } A.$$

So if gas costs \$1.87 per gallon and if gas prices go up by 6%, then the new price of gas will be $(100 + 6)\% = 106\%$ of \$1.87, which is

$$1.06 \times \$1.87 = \$1.98$$

per gallon.

Notice that we can also solve the following type of problem, where the percent increase and the final amount are known, and the initial amount is to be calculated: If the price of gas went up 5% and is now \$2.15 per gallon, then how much did gas cost before this increase? If A was the initial price of gas before the increase, then 105% of A is \$2.15. Therefore

$$1.05 \times A = \$2.15,$$

so

$$A = \$\frac{2.15}{1.05} = \$2.05,$$

so gas cost \$2.05 per gallon before the 5% increase.

The situation is similar when the percent decrease is given. As seen in Figure 3.22, if an amount A decreases by 28% to an amount B, then

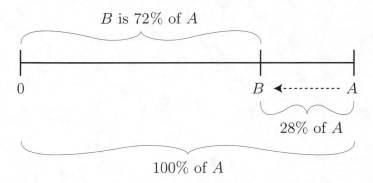

Figure 3.22: Calculating Amounts from the Percent Decrease

$$B \text{ is } 72\% \text{ of } A.$$

In general, if an amount A inceses by $P\%$ to an amount B, then

$$B \text{ is } (100 - P)\% \text{ of } A.$$

So if gas costs $1.87 per gallon and if gas prices go down by 6%, then the new price of gas will be $(100 − 6)\% = 94\%$ of $1.87, which is

$$.94 \times \$1.87 = \$1.76$$

per gallon.

Once again, we can also solve the following type of problem, where the percent decrease and the final amount are known, and the initial amount is to be calculated: If the price of gas went down 5% and is now $2.15 per gallon, then how much did gas cost before this decrease? If A was the initial price of gas before the decrease, then 95% of A is $2.15. Therefore

$$.95 \times A = \$2.15,$$

so

$$A = \$\frac{2.15}{.95} = \$2.26,$$

so gas cost $2.26 per gallon before the 5% decrease.

The Importance of the *Reference Amount*

When you find a percent increase or decrease, make sure you find the increase or decrease as a percent *of the reference amount*, namely the amount for which you want to know the percent increase or decrease of. Let's say the cost of a Dozey-Chair goes from $200 to $300. Then the new price is 50% more than the old price but the old price is only 33% less than the new price. In both percent calculations, the change is $100. The different percentages come about because of the different reference amounts used. In the first case, the reference amount is $200, and $100 is 50% of $200. In the second case, the reference amount is $300, and $100 is only 33% of $300.

The *reference amount* plays the same role in percentages as the *whole* does in fractions: a percent is *of* some quantity, just as a fraction is *of* some whole.

Class Activity 3G: Percent Increase and Decrease

Class Activity 3H: Percent *Of* Versus Percent Increase or Decrease

Exercises for Section 3.5 on Percent Increase and Percent Decrease

1. Last year, Ken had 2.5 tons of sand in his sand pile. This year, Ken has 3.5 tons of sand in his sand pile. By what percent did Ken's sand pile increase from last year to this year? First solve the problem by drawing a picture. Explain how your picture helps you solve the problem. Then solve the problem numerically.

2. Last year's profits were $16 million, but this year's profits are only $6 million. By what percent did profits decrease from last year to this year? First solve the problem by drawing a picture. Explain how your picture helps you solve the problem. Then solve the problem numerically.

3. If sales taxes are 6%, then how much should you charge for an item so that the total cost, including tax, is $35?

4. A pair of shoes has just been reduced from $75.95 to $30.38. Fill in the blanks:

   (a) The new price is ____% less than the old price.

   (b) The new price is ____% of the old price.

   (c) The old price is ____% higher than the new price.

   (d) The old price is ____% of the new price.

5. John bought a piece of land next to land he owns. Now John has 25% more land than he did originally. John plans to give 20% of his new, larger amount of land to his daughter. Once John does this, how much land will John have in comparison to the amount he had originally? Draw a picture to help you solve this. Then solve it numerically, assuming that John starts with 100 acres of land, say.

6. The population of a certain city went up by 2% from 1996 to 1997 and then went down by 2% from 1997 to 1998. By what percent did the

population of the city change from 1996 to 1998? Did the population go up, go down, or stay the same? Make a guess first, then calculate the answer carefully.

Answers to Exercises for Section 3.5 on Percent Increase and Percent Decrease

1. Using a picture: each strip in Figure 3.23 represents 20% of last year's sand pile. Since two additional strips have been added since last year, that is a 40% increase.

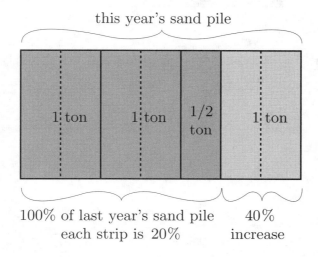

Figure 3.23: Last Year's and This Year's Sand Pile

Calculating the percent increase numerically:

$$3.5 \div 2.5 = 1.4 = 140\%.$$

Subtracting 100%, we see that the percent increase is 40%.

2. Using a picture: each strip in Figure 3.24 represents $2 million. As the picture shows, the decrease in profits is $\frac{1}{8}$ more than 50%. Since $\frac{1}{8} = 12.5\%$, therefore the profits decreased by $50\% + 12.5\%$, namely 62.5%.

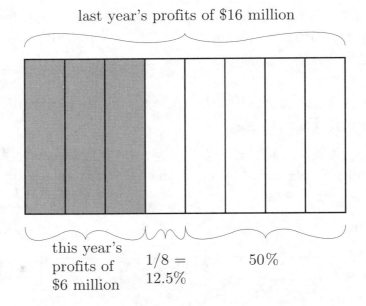

Figure 3.24: Last Year's and This Year's Profits

Calculating the percent decrease numerically:

$$6 \div 16 = .375 = 37.5\%.$$

Therefore the percent decrease is $100\% - 37.5\%$ which is 62.5%.

3. Since the sales tax increases the amount that the customer pays by 6%, therefore if P represents the price of the item, 106% of P must equal 35. Therefore

$$1.06 \times P = \$35,$$

so

$$P = \$35 \div 1.06 = \$33.02.$$

If the price of the item is 33.02, then with a 6% sales tax, the total cost to the customer is 35.

4. (a) The new price is <u>60%</u> less than the old price.

(b) The new price is <u>40%</u> of the old price.

(c) The old price is <u>150%</u> higher than the new price.

(d) The old price is <u>250%</u> of the new price.

5. See Figure 3.25. If John starts with 100 acres of land and gets 25% more, then he will have 125 acres. 20% of 125 acres is 25 acres, so if John gives 20% of his new amount of land away, he will have $125 - 25 = 100$ acres of land, which is the amount he started with.

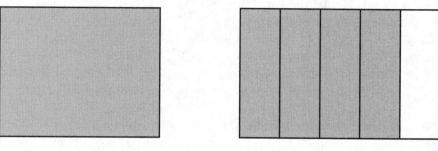

1. John's original plot of land

2. Now John has 25% more land.

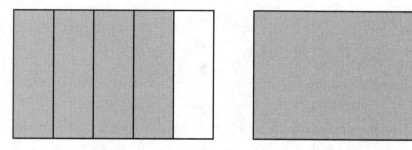

3. 20% of John's new. larger amount of land is represented by the unshaded part.

4. When John gives 20% of his larger plot of land away, he's left with the amount he started with.

Figure 3.25: Solving The Land Problem With A Picture

6. Although you might have guessed that the population stayed the same, it actually decreased by .04%. Try this on a city of 100,000, for example. You can tell that the population will have to decrease because when the population goes back down by 2%, this 2% is *of a larger number* than the 2% increase was of.

Problems for Section 3.5 on Percent Increase and Percent Decrease

1. Last year, the population of South Skratchankle was 48,000, this year the population is 60,000. By what percent did the population increase? First solve the problem by drawing a picture. Explain how your picture helps you solve the problem. Then solve the problem numerically.

2. Last year's sales were $7.5 million. This year's sales are only $6 million. By what percent did sales decrease from last year to this year? First solve the problem by drawing a picture. Explain how your picture helps you solve the problem. Then solve the problem numerically.

3. Jayna and Lisa are comparing the prices of two boxes of cereal of equal weight. Brand A costs $3.29 per box and brand B costs $2.87 per box. Jayna calculates that

$$\frac{\$3.29}{\$2.87} = 1.15.$$

Lisa calculates that

$$\frac{\$2.87}{\$3.29} = .87.$$

   (a) Use Jayna's calculation to make *two* correct statements comparing the prices of brand A and brand B with percentages.

   (b) Use Lisa's calculation to make *two* correct statments comparing the prices of brand A and brand B with percentages.

4. Are the following two problems solved in the same way? Are the answers the same? Solve both and compare how you solved them.

   (a) A television that originally cost $500 is marked down by 25%. What is its new price?

   (b) Last week a store raised the price of a television by 25%. The new price is $500. What was the old price?

5. How much should Swanko Jewelers charge now for a necklace if they want the necklace to cost $79.95 when they reduce their prices by 60%? Explain the reasoning behind your method of calculation.

6. If sales taxes are 7%, then how much should you charge for an item if you want the total cost, including tax, to be $15? Explain the reasoning behind your method of calculation.

7. Connie and Benton bought identical plane tickets, but Benton spent more than Connie (and Connie did not get her ticket for free).

   (a) If Connie spent 25% less than Benton, then did Benton spend 25% more than Connie? If not, then what percent more than Connie did Benton spend?

   (b) If Benton spent 25% more than Connie, then did Connie spend 25% less than Benton? If not, then what percent less than Benton did Connie spend?

8. Of the following 5 statements, which ones have the same meaning? In other words, which of these statements could be used interchangeably (in a news report, for example)? Explain your answers.

   (a) The price went up by 53%.

   (b) The price went up by 153%.

   (c) The new price is 153% of the old price.

   (d) The new price is higher than the old price by 53%.

   (e) The old price was 53% less than the new price.

9. Write a paragraph explaining the difference between a 150% increase in an amount and 150% of an amount. Give examples to illustrate.

10. The SuperDiscount store is planning a "35% off sale" in two weeks. This week, a pair of pants costs $59.95.

    (a) Suppose SuperDiscount raises the price of the pants by 35% this week, and then two weeks from now, lowers the price by 35%. How much will the pants cost two weeks from now? Explain your method of calculation. Explain why it makes sense that the pants won't go back to their original price of $59.95 two weeks from now (a picture may help you explain).

(b) By what percent does SuperDiscount need to raise the price of the pants this week, so that two weeks from now, when they lower the price by 35%, the pants will go back to the original price of $59.95? Explain your method of calculation.

11. Every week, DollarDeals lowers the price of items they have in stock by 10%. Suppose that the price of an item has been lowered twice, each time by 10% of that week's price. Explain why it makes sense that the total discount on the item is *not* 20%, even though the price has been lowered twice by 10% each time. A picture may help you explain. What percent is the total discount?

12. According to the 2000 Census, from 1990 to 2000 the population of Clarke County, Georgia increased by 15.86% and the population of adjacent Oconee County increased by 48.85%.

(a) Can we calculate the percent increase in the total population of the two-county Clarke/Oconee area from 1990 to 2000 by adding 15.86% and 48.85%? Why or why not?

(b) Use the census data in the table below to calculate the percent increase in the total population of the two-county Clarke/Oconee area from 1990 to 2000.

county	1990 population	2000 population
Clarke	87,594	101,489
Oconee	17,618	26,225

13. In 1999, Washington county had a total population—urban and rural populations combined—of 200,000. From 1999 to 2000, the rural population of Washington county went up by 4%, and the urban population of Washington county went up by 8%.

(a) Based on the information above, make a reasonable guess for the percent increase of the total population of Washington county from 1999 to 2000. Based on your guess, what do you expect the total population of Washington county to have been in 2000?

(b) Make up three very different examples for the rural and urban populations of Washington county in 1999, i.e., pick pairs of numbers that add to 200,000. For each example, calculate the total population in 2000, and calculate the percent increase in the total population of Washington county from 1999 to 2000. Compare these answers to your answers in part (a).

(c) If you only had the data given at the beginning of the problem (the 4% and 8% increases and the total population of 200,000 in 1999), would you be able to say exactly what the total population of Washington county was in 2000? Could you give a range for the total population of Washington county in 2000, in other words, could you say that the total population must have been between certain numbers in 2000? If so, what is this range of numbers?

14. One year on *National Equal Pay Day*, news reports stated that on average, women earn 74 cents for every dollar men earn.

(a) How might this "74 cents for every dollar" have been calculated? Make a reasonable guess and describe specifically how to do the calculation. (Different people may have different ideas for this.)

(b) Assuming that the "74 cents for every dollar" statement is true, how much longer must the average woman work than the average man in order to make the same amount of money? Give a "ballpark" answer first, then find the exact answer.

(c) According to the description above, women make 26% less than men on average. Does it follow that the average woman has to work 26% longer in order to make the same amount of money as the average man? What does your answer to part (b) say about this?

3.6   The Commutative and Associative Properties of Addition and Mental Math

In this section, we will study some techniques that can help us solve addition and subtraction problems mentally. Some of these techniques come from basic properties that we assume to be true for addition. These properties of addition are not only helpful for mental calculations, but they are also fundamental to the study of algebra.

Class Activity 3I: Mental Math

In our study of the properties of addition, and later, in our study of other properties of arithmetic, we will consider mathematical expressions involving three or more numbers. Sometimes we will want to group the numbers in an expression in certain ways. In order to do this, we will need to use parentheses.

Parentheses in Expressions with Three or More Terms

A sum

$$A + B + C$$

with 3 terms means the sum of $A + B$ and C. In other words, according to the meaning of $A + B + C$, this sum is to be calculated by first adding A and B, and then adding on C. Likewise, if there are 4 or more terms in a sum, the sum stands for the result obtained by adding from left to right.

But what if we want to indicate the sum of A with $B + C$? We can show this with parentheses. In ordinary writing, parentheses are used for asides, but in mathematical expressions, parentheses[1] are used to group numbers and operations ($+$, $-$, $\times$, $\div$). To indicate the sum of A with $B + C$, simply write

$$A + (B + C).$$

So

$$
\begin{aligned}
17 + (18 + 2) &= 17 + 20 \\
&= 37.
\end{aligned}
$$

Notice that we can express the meaning of the sum

$$A + B + C$$

by using parentheses:

$$(A + B) + C.$$

---

[1]Fortunately, footnotes aren't used any differently.

The Associative Property of Addition

How can you make the problem

$$7384 + 999 + 1$$

easy to solve mentally? Rather than adding from left to right, it is easier to first add 999 and 1 to make 1000, and then add 7384 and 1000 to get 8384. In other words, rather than adding from left to right, calculating

$$(7384 + 999) + 1$$

according to the meaning of the sum $7384 + 999 + 1$, it is easier to group the 999 with the 1 and calculate

$$7384 + (999 + 1)$$

instead. Why is it legitimate to switch the way the numbers in the sum are grouped? Because according to the *associative property of addition*, both ways of calculating the sum give equal results, in other words,

$$(7384 + 999) + 1 = 7384 + (999 + 1).$$

The **associative property of addition** tells us that when we add any three numbers, it doesn't matter whether we add the first two and then add on the third, or we add the first number to the sum of the second and the third—either way we will always get the same answer. In other words, the associative property of addition says that if A, B, and C are any three numbers, then

$$(A + B) + C = A + (B + C),$$

i.e., the sum of $A + B$ and C is equal to the sum of A and $B + C$.

It might help you to remember the terminology *associative property* by remembering that to *associate* with someone means to keep company with them. In

$$(7384 + 999) + 1 = 7384 + (999 + 1),$$

the number 999 can either *associate* with 7384, or it can *associate* with 1.

We assume that the associative property of arithmetic is true for all numbers. But we can use an example to see why this property of addition makes sense. Suppose that Joe has 2 marbles, Sue has 3 marbles, and Terrell has 4 marbles. Sue and Terrell will give all their marbles to Joe, but let's compare two different ways this might happen:

First way: Sue gives her marbles to Joe and then Terrell gives his marbles to Joe.

	Joe	Sue	Terrell
Beginning	oo	ooo	oooo
		Sue's go to Joe	
Middle	oo + ooo		oooo
			Terrell's go to Joe
End	(oo + ooo) + oooo		

According to this way of giving Joe the marbles, Joe has

$$(2+3)+4$$

marbles in all. Why are the parentheses placed this way? At the last step, when Terrell gives Joe his 4 marbles, Joe already has the $2+3$ marbles from Sue, so the 4 marbles are added to the *combined amount* of $2+3$.

Second way: Terrell gives his marbles to Sue and then Sue gives the combined collection of Terrell's and her marbles to Joe.

	Joe	Sue	Terrell
Beginning	oo	ooo	oooo
			Terrell's go to Sue
Middle	oo	ooo + oooo	
		Sue's go to Joe	
End	oo + (ooo + oooo)		

According to this way of giving Joe the marbles, Joe has

$$2+(3+4)$$

marbles in all. Why are the parentheses placed this way? This time Joe gets the *combined amount* of $3+4$ marbles which is added to the 2 marbles that he already has.

Of course, Joe will have the same number of marbles in all, no matter which way he gets the marbles. We can express this by writing

$$(2+3)+4 = 2+(3+4).$$

Of course, the same kind of relationship would hold if other numbers of marbles replaced the numbers 2, 3, and 4. The associative property of addition states that this kind of relationship always holds, no matter what numbers are involved.

The Commutative Property of Addition

How can you make the addition problem

$$2997 + 569 + 3$$

easy to solve mentally? Adding 2997 and 569 is hard, but if we combine the 2997 with the 3 to make 3000, then it's easy to add on the 569 to make 3569. In other words, rather than adding the numbers from left to right in the order they appear, it's easier to switch the order of the 569 and the 3, and calculate

$$2997 + 3 + 569$$

instead. Why is it legitimate to switch the order of 569 and the 3 in the original sum? Because

$$569 + 3 = 3 + 569,$$

and this is true because of the *commutative property of addition.*

The **commutative property of addition** states that for all real numbers A and B,

$$A + B = B + A,$$

in other words, if you have A objects and you get B more objects, then you have the same total number of objects as if you had started with B objects and then got A more objects. To illustrate the commutative property, assume you start with 3 apples and you get 2 more. Then you have

$$3 + 2$$

apples. But what if you had started instead with 2 apples and then got 3 more? In this case you'd have

$$2 + 3$$

apples. Common sense tells us it doesn't matter in which order you got the apples, either way you'll have the same number of apples in all, as we see in Figure 3.26. Another way to say that you have the same amount of apples either way is to say:

$$3 + 2 = 2 + 3.$$

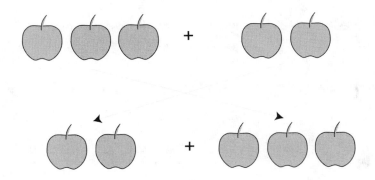

Figure 3.26: $3 + 2 = 2 + 3$

The commutative property of addition says that no matter what the numbers A and B are, even if they are fractions or decimals, if you start with A objects and get B more objects, you have same number in all as if you started with B objects and got A more. In other words:

$$A + B = B + A$$

for all numbers A and B.

In this setting, the word *commute* means *to change places with*. In going from $3 + 2$ to $2 + 3$, the 2 and the 3 change places. The term **turnarounds** is sometimes used with children instead of *commutative property* because it seems to be easier to understand.

Students sometimes get confused between the commutative and associative properties. Notice that the commutative property involves changing the order of *two* terms, while the associative property involves changing the location of parentheses when *three* terms are involved.

Writing Equations That Correspond to a Method of Calculation

We have just seen that the associative and commutative properties of addition can sometimes be used to make mental addition problems easier. In order to show a line of reasoning for a calculation, it is useful to write sequences of equations that correspond to the line of reasoning.

An equation is a mathematical statement saying that two numbers or expressions are equal. The familiar equal sign, $=$, shows this equivalence. Remember that the equal sign means one and only one thing: $=$ means *equals* or *is equal to*.

An equation is actually just a sentence, and you should read it and interpret it as a sentence. The equation

$$7 + (6 + 4) \ = \ 17$$

is the sentence:

7 plus the sum of 6 and 4 is equal to 17.

In order to show a method of calculation, we often string several equations together. Rather than writing

$$198 + 357 + 2 \ = \ 198 + 2 + 357,$$
$$198 + 2 + 357 \ = \ 200 + 357,$$
$$200 + 357 \ = \ 557$$

it is neater and easier to follow this sequence of steps:

$$198 + 357 + 2 \ = \ 198 + 2 + 357$$
$$= \ 200 + 357$$
$$= \ 557$$

When several equations are strung together, it is for the purpose of concluding that the *first* expression written is equal to the *last* expression written. The purpose of the equations above is to conclude that $198 + 357 + 2$ is equal to 557. This uses a general property of equality, which we assume holds true:

If A is equal to B and if B is equal to C, then A is also equal to C.

A Common Error In Using $=$

It is common for students at all levels (including college) to make careless, incorrect use of the equal sign. Because the proper use of the equal sign is essential in algebra, and because elementary school mathematics lays the foundation for learning algebra, it is especially important for you to use the equal sign correctly.

When you use an equal sign, *make sure that the quantities before and after the equal sign really are equal to each other.* It is easy to make a careless error with the equal sign when solving a problem in several steps. Suppose you calculate

$$499 + 165$$

by taking 1 from 165, adding this 1 to 499 to make 500, and then adding on the remaining 164 get 664. How can you show this method of calculation with equations? It is easy to make the following mistake:

$$499 + 1 \; = \; 500 + 164$$
$$= \; 664.$$

This is incorrect because $499 + 1$ *is not equal to* $500 + 164$. Instead, you can write the following correct equations:

$$499 + 165 \; = \; 499 + 1 + 164$$
$$= \; 500 + 164$$
$$= \; 664.$$

Using the Associative and Commutative Properties to Solve Addition Problems

Both children and adults use the associative and commutative properties of addition when solving addition problems mentally. When we use these properties we are often not even aware we are doing so. In fact, is good to develop an easy and fluent use of the properties, because they are essential in algebra.

Children will use the commutative property of addition if they are ready to do so. Young children (first grade or so) often solve addition problems by "counting on." To solve $8 + 2$, a child could start with 8 and count: 9, 10. If the child is given the problem of solving $2 + 8$, the child might start with 2 and count on 8 more. Or the child might use the commutative property— even if he does not know this term—and replace the problem $2 + 8$ with the problem $8 + 2$, and count on from 8 instead of from 2.

Children also use the associative propery of addition spontaneously. When Arianna, a Kindergartener, was asked how to solve $8 + 7$, she described her reasoning this way, refering to a little jingle about $8+8$ that she had learned earlier:

> I know that *eight and eight is sixteen, dip your nose in ice-cream.*
> So 8 and 7 is one less, which is 15.

The following equations correspond to Arianna's method:

$$
\begin{aligned}
8 + 7 &= 8 + (8 - 1) \\
&= (8 + 8) - 1 \\
&= 16 - 1 \\
&= 15.
\end{aligned}
$$

We can see by the way that the parentheses were switched at the second $=$ sign that Arianna used the associative property, even though she certainly didn't know that term. Actually, since the associative property refers to *addition* and not *subtraction*, and since subtracting 1 is the same as adding -1, it is better to use the following equations to demonstrate the use of the associative property:

$$
\begin{aligned}
8 + 7 &= 8 + (8 - 1) \\
&= 8 + (8 + (-1)) \\
&= (8 + 8) + (-1) \\
&= 16 + (-1) \\
&= 15.
\end{aligned}
$$

We can also use the associative property of addition to calculate the sum of two numbers by *shifting* a part of one number to the other number. For example, to solve

$$489 + 73$$

mentally you could use this reasoning:

> If I take 11 of the 73 and add it to 489, the result is 500. This 500 plus the remaining 62 give 562.

You can see why this method makes sense by imagining that you are combining 489 toothpicks with 73 toothpicks. How many toothpicks do you have in all? If you move 11 of the 73 toothpicks (1 ten and 1 one) to the pile of 489

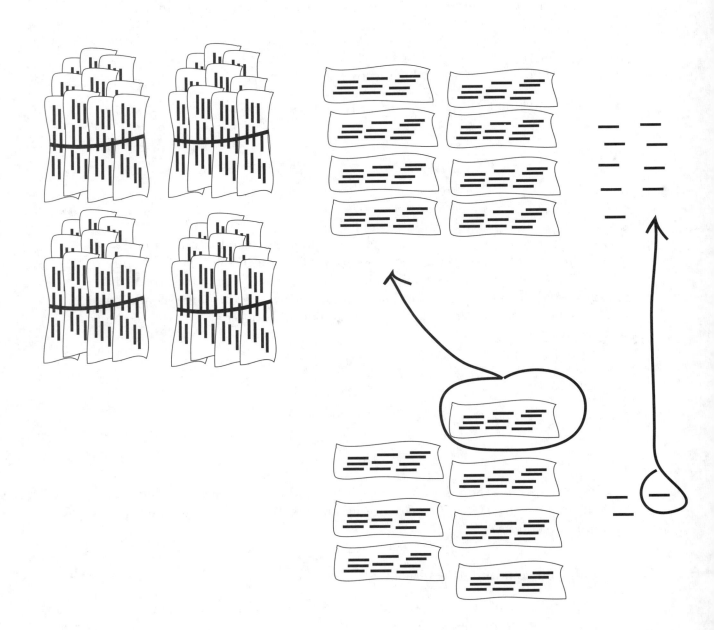

Figure 3.27: Shifting 11 Toothpicks to Change $489 + 73$ into $500 + 62$

toothpicks, then the large pile now has 500 toothpicks, and the small pile has 62 toothpicks. All together that's 562 toothpicks. Figure 3.27 illustrates this moving of toothpicks. When we write equations corresponding to this method of solution, we see that we used the associative property of addition:

$$
\begin{aligned}
489 + 73 &= 489 + (11 + 62) \\
&= (489 + 11) + 62 \\
&= 500 + 62 \\
&= 562
\end{aligned}
$$

The Value of the Commutative and Associative Properties of Addition

The commutative and associative properties of addition allow for great flexibility in calculating sums: you can rearrange the terms any way you like, and you can combine the terms with each other any way you like. This flexibility is valuable not only for mental calculations, but is also an essential skill in algebra. When we combine like terms in the expression

$$3x + 5 + 2x^2 - 4 + 8x - 5x^2,$$

in order to rewrite it as

$$-3x^2 + 11x + 1,$$

we have used the commutative and associative properties of addition.

Class Activity 3J: Using Properties of Addition in Mental Math

Class Activity 3K: Using Properties of Addition to Aid Learning of Basic Addition Facts

Other Mental Methods of Addition and Subtraction

In addition to the associative and commutative properties of addition, there are other ways to use reasoning solve addition and subtraction problems mentally.

Rounding and Compensating

Sometimes an addition or subtraction problem involves a number that is close to a round number that is easy to work with. In this case, we can either use the associative property to "shift" numbers (such as in Figure 3.27), or we can round the numbers, and compensate for that rounding after adding or subtracting.

To solve $376 + 199$ we can round and compensate:

> Suppose we add 200 to 376 instead of adding 199 to 376. This makes 576. But we added 1 too many, so we must take 1 away from 576. Therefore $376 + 199 = 575$.

We can write corresponding equations as follows:

$$
\begin{aligned}
376 + 199 &= 376 + 200 - 1 \\
&= 576 - 1 \\
&= 575.
\end{aligned}
$$

We can also round and compensate in subtraction problems. Consider the problem $684 - 295$. The number 295 is close to 300. So to calculate

$$684 - 295$$

we can reason as follows:

> Taking 300 objects away from 684 objects leaves 384 objects. But when we took 300 away, we took away 5 more than we should have (because 300 is 5 more than 295). This means that we must *add* 5 to 384 to get the answer, 389.

The equations below correspond to this line of reasoning.

$$
\begin{aligned}
684 - 295 &= 684 - 300 + 5 \\
&= 384 + 5 \\
&= 389
\end{aligned}
$$

Subtracting by Adding On

Another way to solve $684 - 295$ is to view the difference $684 - 295$ as the amount we must add to 295 to make 684. This interprets subtraction as *comparison*, as was discussed in Section 3.1. With this point of view, we start with 295 and keep adding on numbers until we reach 684:

$$\begin{aligned} 295 + 5 &= 300 \\ 300 + 300 &= 600 \\ 600 + 84 &= 684, \end{aligned}$$

or

$$295 + 5 + 300 + 84 = 684.$$

All together, starting with 295, we added on

$$5 + 300 + 84 = 389,$$

to reach 684. Therefore $684 - 295 = 389$. This is the method that shopkeepers frequently use to make change.

Why Should We be Able to Use Reasoning to Add and Subtract?

Above, we saw how to use the associative and commutative properties and other ways of reasoning in order to solve addition and subtraction problems. Why should we be able to use these methods when we already have standard algorithms that always work? Although it might be easier to teach only the standard algorithms, the non-standard methods encourage reasoning that is essential in the study of algebra. The mathematics that you will teach in elementary school will lay the foundation for algebra. This foundation must be strong, so that your students will be able to progress in mathematics. Currently, most high schools require at least two years of math to graduate.

A strong foundation in mathematics will be a wise economic investment for your students. According to the Bureau of Labor Statistics of the U.S. Department of Labor, on average, workers with more education have higher earnings and a lower unemployment rate (see [37]). The U.S. Department of Education's paper "Mathematics Equals Opportunity" (see [36]) reported that students who take rigorous mathematics courses are more likely to go on to college, that algebra is a "gateway" course to more advanced mathematics courses in high school, and that low-income students who take Algebra I and Geometry are much more likely to attend college than those who do not. It is not an exaggeration to say that a strong foundation in mathematics can have a lifelong positive impact.

In addition, according to the National Center for Education Statistics, workers who had scored in the top quartile on the Armed Services Vocational Aptitude Battery test in mathematics earned on average over 35% more per hour than workers with lower scores (see [17]). Some of the fastest growing jobs, such as those involving computers or health care, often require a good ability with math. Some jobs that seem like they wouldn't need math actually do. For example, according to the Bureau of Labor Statistics (see [38]) foresters and conservation scientists require advanced levels of theoretical math. Landscape architects, aircraft pilots, and veterinarians need to know applied math. Sheet-metal workers, bricklayers, and jewelers need to know practical "shop" math, and arithmetic is important for ticket agents and stock clerks. Of course, there are many excellent jobs that don't require much math, but education should equip students with skills that prepare them for a broad range of opportunities.

Class Activity 3L: Writing Correct Equations

Class Activity 3M: Writing Equations that Correspond to a Method of Calculation

Class Activity 3N: Rounding and Compensating

Class Activity 3O: Other Ways to Add and Subtract

Class Activity 3P: A Third Grader's Method of Subtraction

Exercises for Section 3.6 on the Commutative and Associative Properties of Addition and Mental Math

1. Give an example that demonstrates how the associative property of addition can be used to make a problem easier to do mentally. Show your use of the associative property explicitly.

2. The sequence of equations below shows a way of using properties of addition to calculate a sum. Say specifically which properties of addition were used, and where.

$$
\begin{aligned}
27 + 89 + 13 &= 27 + (89 + 13) \\
&= 27 + (13 + 89) \\
&= (27 + 13) + 89 \\
&= 40 + 89 \\
&= 129
\end{aligned}
$$

3. For each of the following addition problems, write equations that correspond to a mental method for calculating the sum that uses the associative and/or the commutative properties of addition. Say specifically which properties of addition were used, and where.

   (a) $993 + 2389$

   (b) $398 + (76 + 2)$

4. Each arithmetic problem below has a description for how to solve the problem. In each case, write a sequence of equations that correspond to the given description.

   (a) Problem: $23 + 45$

   Solution: 23 plus 40 is 63, and then 5 more makes 68.

   (b) Problem: $800 - 297$

   Solution: $800 - 300 = 500$, but subtracting 300 subtracts 3 too many, so we must add 3 back, making 503.

5. Nancy writes the following equations in order to solve $37 + 14$:

$$30 + 10 = 40 + 7 = 47 + 4 = 51.$$

   Write correct equations that solve $37 + 14$ and that incorporate Nancy's solution strategy.

6. Jim writes the following equations in order to solve $85 - 15$:

$$85 - 10 = 75 - 5 = 70.$$

   Write correct equations that solve $85 - 15$ and that incorporate Jim's solution strategy.

7. Find ways to solve the following addition and subtraction problems *other than* using the standard addition or subtraction algorithms. In each case explain your reasoning and also write equations that incorporate your thinking.

    (a) $786 - 47$

    (b) $427 + 28$

    (c) $999 + 999$

    (d) $1002 - 986$

    (e) $237 - 40$

Answers to Exercises for Section 3.6 on the Commutative and Associative Properties of Addition and Mental Math

1. To calculate $49 + 37$ mentally, we can think of moving 1 from the 37 to the 49, so that the sum becomes $50 + 36$, which we can easily see is 86. The corresponding equations show that the associative property of addition was used:

$$
\begin{aligned}
49 + 37 &= 49 + (1 + 36) \\
&= (49 + 1) + 36 \\
&= 50 + 36 \\
&= 86
\end{aligned}
$$

    The associative property of addition is used at the 2nd equal sign to switch the placement of parentheses from grouping 1 and 36 together, to grouping 49 and 1 together.

2. The associative property of addition was used at the first $=$ sign to say that $(27 + 89) + 13 = 27 + (89 + 13)$. The commutative property of addition was used at the second $=$ sign to change $89 + 13$ to $13 + 89$. The associative property of addition was used at the third $=$ to say that $27 + (13 + 89) = (27 + 13) + 89$.

3. (a) $993 + 2389 = 993 + (7 + 2382) = (993 + 7) + 2382 = 1000 + 2382 = 3382$. The associative property of addition was used in rewriting $993 + (7 + 2382)$ as $(993 + 7) + 2382$.

(b) $398 + (76 + 2) = 398 + (2 + 76) = (398 + 2) + 76 = 400 + 76 = 476$.
The commutative property of addition was used to rewrite $76 + 2$
as $2 + 76$. The associative property of addition was used to rewrite
$398 + (2 + 76)$ as $(398 + 2) + 76$

4. (a)

$$
\begin{aligned}
23 + 45 &= 23 + (40 + 5) \\
&= (23 + 40) + 5 \\
&= 63 + 5 \\
&= 68.
\end{aligned}
$$

Note that the associative property of addition is used at the second
equal sign in order to switch the placement parentheses. This has
the effect of grouping the 40 with the 23 instead of with the 5
(where it was part of 45).

(b)

$$
\begin{aligned}
800 - 297 &= 800 - 300 + 3 \\
&= 500 + 3 \\
&= 503.
\end{aligned}
$$

5.

$$
\begin{aligned}
37 + 14 &= 30 + 10 + 7 + 4 \\
&= 40 + 7 + 4 \\
&= 47 + 4 \\
&= 51
\end{aligned}
$$

6.

$$
\begin{aligned}
85 - 15 &= 85 - 10 - 5 \\
&= 75 - 5 \\
&= 70
\end{aligned}
$$

7. Here are equations you could write:

(a)

$$786 - 47 \; = \; 786 - 50 + 3$$
$$= \; 736 + 3$$
$$= \; 739$$

(b)

$$427 + 28 \; = \; 427 + 3 + 25$$
$$= \; 430 + 25$$
$$= \; 455$$

(c)

$$999 + 999 \; = \; 999 + 1000 - 1$$
$$= \; 1999 - 1$$
$$= \; 1998$$

(d)

$$986 + 4 \; = \; 990$$
$$990 + 10 \; = \; 1000$$
$$1000 + 2 \; = \; 1002$$
so
$$986 + 4 + 10 + 2 \; = \; 1002$$
so
$$1002 - 986 \; = \; 4 + 10 + 2$$
$$= \; 16$$

(e)

$$237 - 40 \; = \; 240 - 40 - 3$$
$$= \; 200 - 3$$
$$= \; 197$$

Problems for Section 3.6 on the Commutative and Associative Properties of Addition and Mental Math

1. Many teachers have a collection of small cubes that can be snapped end-to-end to make "trains" of cubes. The cubes come in an array of colors.

   (a) Describe how you could use snap-together cubes in different colors to demonstrate the commutative property of addition.

   (b) Describe how you could use snap-together cubes in different colors to demonstrate the associative property of addition.

2. Suppose a child has learned the following:

   - all the sums of whole numbers that add to 10 or less;
   - $10 + 1, \ 10 + 2, \ 10 + 3, \ldots, 10 + 10$;
   - the *doubles* $1 + 1, \ 2 + 2, \ 3 + 3, \ldots, 10 + 10$.

   For each sum below, show at least three *different* ways to use the addition facts the child knows, and the commutative and associative properties, to calculate the sum. In each case, describe how to show the method of calculation by joining a group of white snap-together cubes with a group of red snap-together cubes. (See Problem 1 for more about these kinds of cubes.)

   (a) $6 + 7$

   (b) $8 + 9$ (Try to find four or five different ways to solve this.)

3. To solve $159 - 73$, a student writes the following equations:
   $$160 - 70 = 90 - 3 = 87 - 1 = 86.$$

   Although the student has a good idea for solving the problem, his equations are not correct. In words, describe the student's solution strategy; then write a correct sequence of equations that correspond to this solution strategy. Write your equations in the following form:

   $$
   \begin{aligned}
   159 - 73 &= \text{some expression} \\
   &= \text{some expression} \\
   &= \vdots \\
   &= 86
   \end{aligned}
   $$

4. *Problem:* If I have 12 eggs and use 5 in a recipe, how many eggs will I have left?

   Arianna's solution: There are 7 eggs left. Because if I take 5 away from 10 there will be 5 left. So then I counted 6 for 11 and 7 for 12.

   Write a coherent sequence of equations that solve $12 - 5$ and that show why Arianna's reasoning is valid.

5. Describe a way to solve $304 - 81$ mentally, by using reasoning other than the standard subtraction algorithm. Then write a coherent sequence of equations that correspond to your reasoning.

6. David and Ashley want to solve $8.27 - 2.98$ by first solving $8.27 - 3 = 5.27$. David says that they must *subtract* .02 from 5.27, but Ashley says that they must *add* .02 to 5.27. Who is right and why? Use reasoning to explain which one is correct. Do not just say which answer is numerically correct, give an explanation in words for why the answer must be correct.

7. Tylishia says that she can solve $324 - 197$ by adding 3 to both numbers and solving $327 - 200$ instead. Is Tylishia's method valid? Explain why or why not. Do not just say whether or not Tylishia gets the correct answer using her method, discuss whether or not it is a valid method, and why.

8. Is there an *associative property of subtraction*? In other words, is

$$A - (B - C) = (A - B) - C$$

   true for all real numbers A, B, C? Explain your answer. If the equation is always true, explain why; if the equation is not always true, find another expression that *is* equal to the expression

$$A - (B - C),$$

   and explain why the two expressions are equal.

9. Is there a *commutative property of subtraction*? Explain why or why not. In your answer, include a statement of what a *commutative property of subtraction* would be if there were such a property.

Chapter 4

Multiplication

In this chapter we will study multiplication of all different kinds of numbers: whole numbers, rational numbers, real numbers, and integers. Because multiplication is used so often, it is easy to take the concept for granted, and it is easy to think that the notion is trivial. However, even though it is easy for adults to carry out the *procedure* of multiplying, the underlying *concept* of multiplication is much more subtle. In this chapter, we will study the concept of multiplication, especially the meaning of multiplication, ways of representing multiplication, and properties of multiplication. We will see that the *meaning* of multiplication gives rise to the familiar procedures that we use when we multiply whole numbers, fractions, and decimals.

4.1   The Meaning of Multiplication and Ways to Show Multiplication

In this section we will study the meaning of multiplication. We will also study various ways of picturing and thinking about multiplication of whole numbers; some people call these **multiplication models**. These will provide us with different ways to organize information to make the multiplicative structure in a situation visible. By working with these different ways of displaying multiplication in a variety of situations, you will deepen your understanding of the concept of multiplication. *Part of having a deep understanding of multiplication is to know not only* that *you multiply to solve a certain problem, but to know* why *the problem calls for multiplication, and not some other operation, such as addition.* In order to do this, you will need to focus on the meaning of multiplication.

The Meaning of Multiplication

What does multiplication mean? If A and B are non-negative numbers, then

$$A \times B \quad \text{or} \quad A \cdot B,$$

which we read as "A times B", means the total number of objects in A groups if there are B objects in *each* group. The result $A \times B$ is called the **product** of A and B, and the numbers A and B are called **factors**.   So if there are 237 bags and each bag contains 46 potatoes, then the total number of potatoes is 237×46, which turns out to be $10,902$ potatoes.

A whole number A that is whole number times a whole number B is called a **multiple** of B. For example, 15 is a multiple of 5 and the numbers

$$4, \ 8, \ 12, \ 16, \ 20, \ 24, \ldots$$

are multiples of 4.

Showing Multiplicative Structure by Grouping

The simplest way to show that a situation requires multiplication is to use **grouping**. When a collection of objects is arranged into A groups with B objects in each group then we know that *according to the meaning of multiplication*, there are $A \times B$ objects in all. Figure 4.1 shows that there are a total of

$$3 \times 4$$

dots pictured because the dots are arranged into 3 groups of 4.

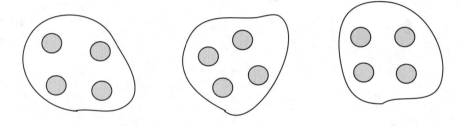

Figure 4.1: Grouped Objects Show Multiplicative Structure

Showing Multiplicative Structure with Organized Lists

Another way to show that a situation involves multiplication is with an **organized list**. Some examples will illustrate. Suppose you want to represent the different ways to order a burger and fries at a restaurant. Let's say there are two types of burgers to choose from: cheeseburger C, regular burger B; and let's say there are three types of fries to choose from: small S, large L, jumbo J. Then we can make the following organized list to show all the possibilities for the combination of a burger and fries:

B S	meaning regular burger and small fries
B L	meaning regular burger and large fries
B J	meaning regular burger and jumbo fries
C S	meaning cheeseburger and small fries
C L	meaning cheeseburger and large fries
C J	meaning cheeseburger and jumbo fries

This way of organizing the information shows *why* we can multiply to determine the number of choices for the combination of a burger and fries. As Figure 4.2 shows, the organized list consists of 2 natural groups with 3 items in each group (where each item in the group consists of a *combination* of burger and fries). Therefore, *according to the meaning of multiplication*, there are

$$2 \times 3$$

ways to choose the combination of a burger and fries.

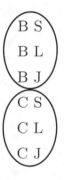

Figure 4.2: An Organized List Shows Multiplicative Structure

Now suppose that in addition to the two type of burgers (C, B) and the three types of fries (S, L, J), you also have a choice of two types of cola: regular (R) and diet (D). How many different ways are there to get a combination of a cola, a burger, and fries? Once again, we can show all possible combinations with an organized list:

R B S   meaning regular cola, regular burger and small fries
R B L   meaning regular cola, regular burger and large fries
R B J
R C S
R C L
R C J   meaning regular cola, cheeseburger and jumbo fries
D B S
D B L   meaning diet cola, regular burger and large fries
D B J
D C S
D C L
D C J   meaning diet cola, cheeseburger and jumbo fries

As Figure 4.3 shows, this organized list consists of 2 natural groups. When we look in closer detail, we see that each of the 2 groups can be broken into 2 smaller groups, with 3 entries in each of these smaller groups. Therefore, *according to the meaning of multiplication*, there are

$$2 \times (2 \times 3)$$

ways to choose the combination of a cola, a burger, and fries, which works out to 12 different ways. Notice that the parentheses in the expression $2 \times (2 \times 3)$ are used to show that there are 2 large groups that each contain 2×3 items.

Showing Multiplicative Structure with Array Diagrams

Consider once again the situation of two types of burgers (regular burger B, and cheeseburger C) and three types of fries (small S, large L, and jumbo J) to choose from at a restaurant. How many ways are there to choose the combination of a burger and fries? We can use an **array diagram** to display all the ways of choosing a burger and fries:

B, S	B, L	B, J
C, S	C, L	C, J

Each entry in the array corresponds to a choice of burger and fries. For example, the entry C, L corresponds to the choice of a cheeseburger and large fries. If we consider each row in the array to be a group, then the array

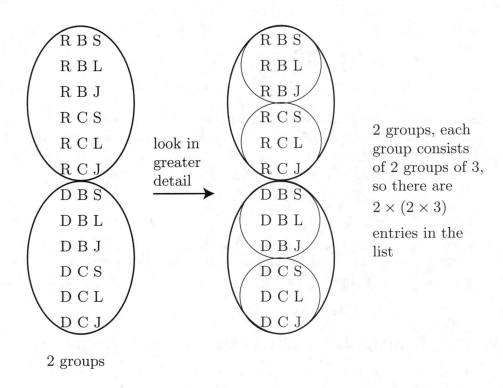

Figure 4.3: An Organized List Shows Multiplicative Structure

consists of 2 groups with 3 items in each group. Therefore, *according to the meaning of multiplication*, there are

$$2 \times 3$$

items in the array, and thus 2×3 ways to choose the combination of a burger and fries. Notice that if we choose our groups to be the *columns* of the array rather than the rows, then we would say that there are

$$3 \times 2$$

items in the array, because in this case there are 3 groups of 2.

Now consider the situation of two types of cola, regular cola R, and diet cola D, in addition to the two types of burgers (B, C) and three types of fries (S, L, J). How many ways are there to choose the combination of cola, a burger, and fries? We can show all possible choices by showing *two* array diagrams, one for each type of cola:

REGULAR COLA

B, S	B, L	B, J
C, S	C, L	C, J

DIET COLA

B, S	B, L	B, J
C, S	C, L	C, J

Once again, we can see the multiplicative structure in this situation because the choices for a cola, burger, and fries are divided into 2 groups with 2×3 items in each group. Therefore, *according to the meaning of multiplication*, there are

$$2 \times (2 \times 3)$$

ways of choosing the combination of a cola, burger and fries.

Showing Multiplicative Structure with Tree Diagrams

Consider once again the situation of two types of cola (regular cola R, and diet cola D), two types of burgers (regular burger B, and cheeseburger C) and three types of fries (small S, large L, and jumbo J) to choose from at a restaurant. How many ways are there to choose the combination of a cola, a burger, and fries? Before we see how to organize these choices into a tree diagram, consider first the following modification of the organized list that is shown on page 209:

	regular burger	small fries
		large fries
regular cola		jumbo fries
	cheeseburger	small fries
		large fries
		jumbo fries
	regular burger	small fries
		large fries
diet cola		jumbo fries
	cheeseburger	small fries
		large fries
		jumbo fries

When we further modify the way this information is displayed we obtain the **tree diagram** pictured in Figure 4.4.

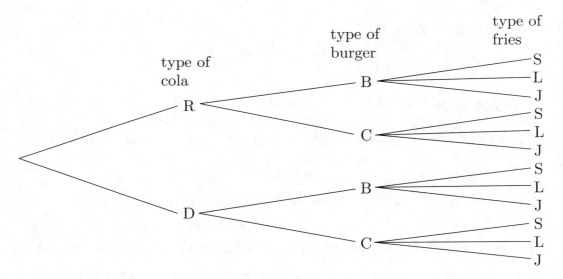

Figure 4.4: A Tree Diagram

To "read" the tree diagram, start at the far left and follow branches *all the way across*. The sequence of branches that you travel along going across corresponds to a choice of a soda, burger and fries. For example, starting at the left, you might pick the R (regular cola) branch, followed by the C (cheeseburger) branch, and then followed by the L (large fries) branch, as shown in Figure 4.5. This choice of branches all the way across corresponds to the entry RCL in the organized list on page 209.

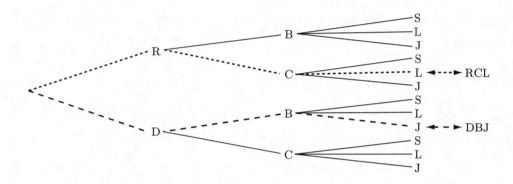

Figure 4.5: Reading A Tree Diagram

As with the other ways of displaying this information, the tree diagram shows multiplicative structure. There are 2 primary branches (R and D), and each of those two primary branches has 2 secondary branches (B and C). Therefore there are 2 groups of 2 secondary branches, so *according to the meaning of multiplication*, there are 2×2 secondary branches. For each of the 2×2 secondary branches, there are 3 third-level (tertiary) branches (S, L, and J). Therefore there are 2×2 groups of 3 third-level branches, and so, *according to the meaning of multiplication*, there are

$$(2 \times 2) \times 3$$

third-level branches. Since each third-level branch corresponds to a choice of cola, burger, and fries, there are also $(2 \times 2) \times 3 = 12$ choices for the combination of a cola, burger, and fries.

Using Organized Lists, Array Diagrams, and Tree Diagrams

Even though organized lists, arrays, and tree diagrams are relatively efficient ways to record data, they can become tedious to draw in their entirety when many numbers or large numbers are involved. However, when many numbers or large numbers are involved, you can often show the multiplicative structure in a situation by *starting* an organized list or tree diagram, and drawing only a portion of it. Even if you can't finish the diagram completely, if you show the *structure* of the full diagram you will be able to explain why multiplication is the appropriate operation to use in the situation.

Class Activity 4A: Showing Multiplicative Structure

Exercises For Section 4.1 on the Meaning of Multiplication

1. How many different three digit numbers can you make using only the digits 1, 2, and 3 if you do not repeat any digits (so that 121 and 332 are not counted)? Show how to solve this problem with an organized list and with a tree diagram. Use the meaning of multiplication to explain why this problem can be solved by multiplying.

2. How many different three-digit numbers can be made using only the digits 1, 2 and 3 where repeated digits *are* allowed (so that 121 and 332 *are* counted)? Show how to solve this problem with an organized list and with a tree diagram. Use the meaning of multiplication to explain why this problem can be solved by multiplying.

3. How many different keys can be made if there are ten places along the key that will be notched and if each notch will be one of eight depths?

4. Annette buys a wardrobe of 3 skirts, 3 pants, 5 shirts and 3 sweaters, all of which are coordinated so that she can mix and match them any way she likes. How many different outfits can Annette create from this wardrobe? (Every day Annette wears either a skirt or pants, a shirt, and a sweater.)

5. A delicatessen offers four types of bread, twenty types of meats, fifteen types of cheese, a choice of mustard, mayonnaise, both or neither, and

a choice of lettuce, tomato, both or neither for their sandwiches. How many different types of sandwiches can the deli make with one meat and one cheese? Should the deli's advertisement read: "hundreds of sandwiches to choose from," or would it be better to substitute *thousands* or even *millions* for *hundreds*?

6. A pizza parlor offers ten toppings to choose from. How many ways are there to order a large pizza with exactly two different toppings? (For example, you could order pepperoni and mushroom, but not double pepperoni.)

Answers to Exercises for Section 4.1 on the Meaning of Multiplication

1. Figure 4.6 shows a tree diagram and an organized list displaying all such 3-digit numbers. Both show a structure of 3 big groups, one group for each possibility for the first digit. Once a first digit has been chosen, there are two choices for the second digit (since repeats aren't allowed). This means there are 3 groups of 2 possibilities for the first two digits. So by the meaning of multiplication, there are $3 \times 2 = 6$ possibilities for the first two digits. Once the first two digits are chosen, there is no choice for the third digit—it must be the remaining unused digit. So there are $3 \times 2 = 6$ three-digit numbers using only the digits 1, 2, 3, with no digit repeated.

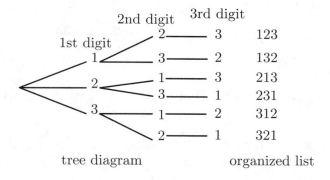

Figure 4.6: Tree Diagram And Organized List

2. The tree diagram and the organized list in Figure 4.7 both show 3 big groups. Each of those 3 big groups has 3 groups of 3. Therefore, according to the meaning of multiplication, there are

$$3 \times (3 \times 3) = 27$$

three-digit numbers that use only the digits 1, 2, and 3.

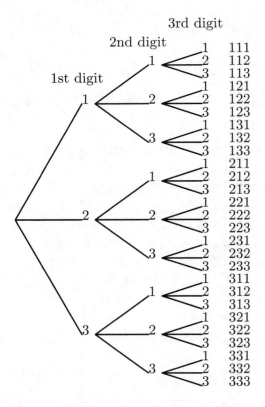

Figure 4.7: Tree Diagram and Organized List

3. Figure 4.8 shows part of a tree diagram for all such keys. There are 8 primary branches corresponding to the 8 choices for the first notch. For each of those 8 choices on the first notch there are 8 choices for the second notch. Therefore there are 8 groups of 8 choices for the first two notches, and so, by the meaning of multiplication, there are 8×8 choices for the first two notches. For each of the 8×8 choices for the

first two notches there are 8 choices for the 3rd notch. Therefore there are $8 \times 8 \times 8$ choices for the first three notches. And so on. When all 10 notches are taken into account, there are $8 \times 8 \times 8 \times 8 \times 8 \times 8 \times 8 \times 8 \times 8 \times 8 = 8^{10} = 1,073,741,824$ that can be made, which is more than a billion keys!

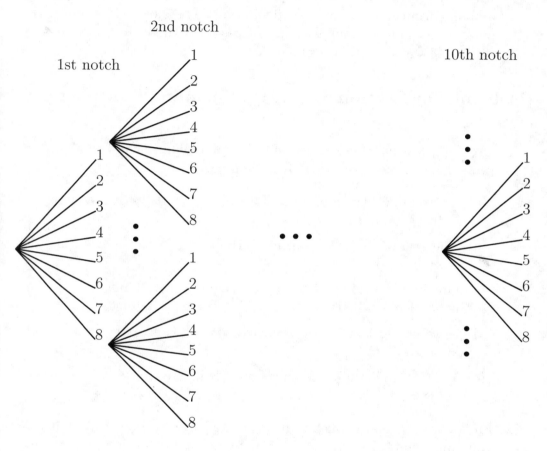

Figure 4.8: A Partial Tree Diagram for Keys

4. Annette can make $6 \times 5 \times 3 = 90$ different outfits.

5. There are $4 \times 20 \times 15 \times 4 \times 4 = 19,200$ different sandwiches that can be made with a meat and a cheese. The deli should substitute *thousands* for *hundreds*.

6. This is just like the *clinking glasses* problem of Chapter 1. For each of the 10 choices for the first topping, there are 9 choices for the second topping (there are only 9 because repeats are not allowed). This forms 10 groups of 9, so there are 10×9 choices of double toppings. However, when they are counted this way, pepperoni and mushrooms counts as different from mushrooms and pepperoni. Therefore we should divide by 2 so as not to double count pizzas. So there are actually only $\frac{10 \times 9}{2} = 45$ ways to order a pizza with two different toppings.

Problems for Section 4.1 on the Meaning of Multiplication

1. Use the *meaning of multiplication* to explain why each of the following problems can be solved by multiplying.

   (a) There are 3 feet in a yard. If a rug is 5 yards long, how long is it in feet?

   (b) There are 5280 feet in a mile. How long in feet is a 4 mile long stretch of road?

   (c) One inch is 2.54 centimeters. Hillary is 5 feet, 3 inches tall, which is 63 inches tall. How tall is Hillary in centimeters?

   (d) A kilometer is .6 miles (approximately). How many miles is a 10 kilometer run?

   (e) Will is driving 65 miles per hour. If he stays at that speed, how far will he drive in $2\frac{1}{2}$ hours?

2. Find a tree or a large bush that has more leaves (or needles) on it than you can count. Use multiplication to estimate the number of leaves (or needles) on the tree. Describe your method and use the meaning of multiplication to explain why the problem can be solved by multiplying.

3. John, Trey, and Miles want to know how many two-letter acronyms there are that don't have a repeated letter. For example, they want to count acronyms such as BA and AT, but they don't want to count acronyms such as ZZ or XX. John says there are $26 + 25$ because you don't want to use the same letter twice, that's why the second number is 25. Trey says he thinks it should be *times*, not *plus*: 26×25. Miles

says the number is $26 \times 26 - 26$ because you need to take away the double letters. Discuss the boys' ideas. Which answers are correct, which are not, and *why*? Explain your answers clearly and thoroughly, drawing on the meaning of multiplication.

4. Allie and Betty want to know how many three-letter acronyms, such as BMW, or DDT are possible (letters are allowed to repeat, as in DDT or BOB). Allie thinks there can be $26 + 26 + 26$ three-letter acronyms while Betty thinks the number is $26 \times 26 \times 26$. Which girl, if either, is right and why? Explain your answers clearly and thoroughly, drawing on the meaning of multiplication.

5. Explain your answers to the following.

   (a) How many nine-digit numbers are there that use only the digits 1, 2, 3, ..., 8, 9 (repetitions allowed, so, for example 123211114 is allowed)?

   (b) How many nine-digit numbers are there that use each of the digits 1, 2, 3, ..., 8, 9 exactly once?

6. In all three parts below, explain your method so that someone who is not in your course could understand why your answer is correct. In particular, if you use an operation such as multiplication or addition, explain why you use *that* operation as opposed to some other operation.

   (a) How many whole numbers are there that have exactly 10 digits and that can be written using only the digits 8 and 9?

   (b) How many whole numbers are there that have at most 10 digits and that can be written using only the digits 0 and 1? (Is this related to part (a)?)

   (c) How many whole numbers are there that have exactly 10 digits and that can be written using only the digits 0 and 1?

7. Most Georgia car license plates currently use the format of three numbers followed by three letters (such as 438 GGT). How many different license plates can be made this way? Explain your method so that someone who is not in your course could understand why your answer is correct. In particular, if you use an operation such as multiplication

or addition, explain why you use *that* operation as opposed to some other operation.

8. (a) How many ways are there of picking a pair of co-presidents from a club of 40 people?

   (b) How many ways are there of picking a president and vice-president from a club of 40 people?

   (c) Are the answers to parts a and b the same or different? Explain why they are the same or why they are different.

9. A dance club has ten women and ten men. In both parts below, give a clear enough explanation so that someone who is not in your course could understand why your method gives the correct answer.

   (a) How many ways are there to choose one woman and one man to demonstrate a dance step?

   (b) How many ways are there to pair each women with a man (so that all ten women and all ten men have a dance partner at the same time)?

10. *A pizza parlor problem.* How many ways are there to order a large pizza with no double toppings? (For example, mushroom, pepperoni, and sausage is one possibility, so is mushroom and sausage, but not double mushroom and sausage. Your count should also include the case of no toppings.) You can't answer the question yet because you don't know how many toppings the pizza parlor offers.

    (a) Answer the question above when there are exactly three toppings to choose from (such as mushroom, pepperoni and sausage).

    (b) Answer the question above when there are exactly four toppings to choose from.

    (c) Answer the question above when there are exactly five toppings to choose from.

    (d) Look for a pattern in your answers in parts (a), (b), (c). Based on the pattern you see, predict the answer to the question above when there are ten toppings to choose from.

(e) Now find a different way to answer the question when there are ten toppings to choose from. This time think about the situation in the following way: pepperoni can be either on or off, mushrooms can be either on or off, sausage can be either on or off, and so on, for all ten toppings. Explain clearly how to use this idea to answer the question, and why this method is valid.

11. A pizza parlor offers ten different toppings to choose from. How many ways are there to order a large pizza if double toppings, but not triple toppings or more are allowed? For example: double pepperoni and (single) mushroom is one possibility, so is double pepperoni and double mushroom, but triple pepperoni is not allowed. Notice that each topping can be chosen in one of three ways: off, as a single topping, as a double topping.

12. Make a tree diagram and a corresponding organized list to show all the ways of making 31 cents out of quarters, dimes, nickels, and pennies. Unlike the other tree diagrams in this section, this tree diagram won't have the same number of branches coming from each branch at the same stage, and therefore won't have a multiplicative structure.

13. Make a tree diagram and a corresponding organized list to show all the ways of presenting 234 toothpicks in bundles of hundreds, bundles of tens, and individual toothpicks. So for example, 2 bundles of 100 toothpicks, 3 bundles of 10 toothpicks, and 4 individual toothpicks is one way to show 234 toothpicks, but so is 2 bundles of 100 toothpicks and 34 individual toothpicks. As in the previous problem, this tree diagram won't have the same number of branches coming from each branch at the same stage, and therefore won't have a multiplicative structure.

4.2 Why Multiplying Decimals by 10 is Easy

Why is multiplying by 10 and by powers of 10 so easy in the decimal system? To multiply a number that is written as a decimal by 10, all we do is *move every digit one place to the left*. We can also think of this as moving the decimal point one place to the right because this has the same effect. Similarly, to multiply by 100, move every digit *two places* to the left. To multiply by 1000, move every digit *three places* to the left, and so on.

Multiplying by 10 and by powers of 10 is so easy because of the special structure of the decimal system: in decimals, the value of each place is 10 times the value of the place to its immediate right. Consider what happens when we multiply the number 34 by 10. The number 34 stands for 3 tens and 4 ones, and can be represented by 3 bundles of 10 toothpicks and 4 individual toothpicks, as shown in Figure 4.9. Then 10×34 stands for the total number

Figure 4.9: 34 Represented by 3 Bundles of 10 and 4 Individual Toothpicks

of toothpicks in 10 groups of 34 toothpicks. As Figure 4.10 shows, when we form 10 groups of 34 toothpicks, each of the 3 original groups of 10 toothpicks becomes bundled into 1 group of 100. Therefore, when we multiply by 10, the 3 in the 10s place moves one place over to the 100s place. *Notice that this shifting occurs precisely because the value of the hundreds place is 10 times the value of the tens place.* Similarly, when we multiply 34 by 10, each of the 4 original individual toothpicks is bundled into 1 group of 10. Therefore, when we multiply by 10, the 4 in the 1s place moves one place over to the 10s place. Once again, *notice that this shifting occurs precisely because the value of the tens place is 10 times the value of the ones place.*

The situation is similar for other decimals. We can think of multiplying

$$10 \times 400,000.007$$

as forming 10 groups of bundled objects representing $400,000.007$. As above:

the 4 hundred-thousand moves one place to the left to become 4 million *because* the value of the millions place is 10 times the value of the hundred-thousands place,

the 7 thousandths move one place to the left to become 7 hundredths *because* the value of the hundredths place is 10 times the value of the thousandths place.

What about multiplying by 100, or 1000, or 10000, and so on? Because

$$100 = 10 \times 10,$$

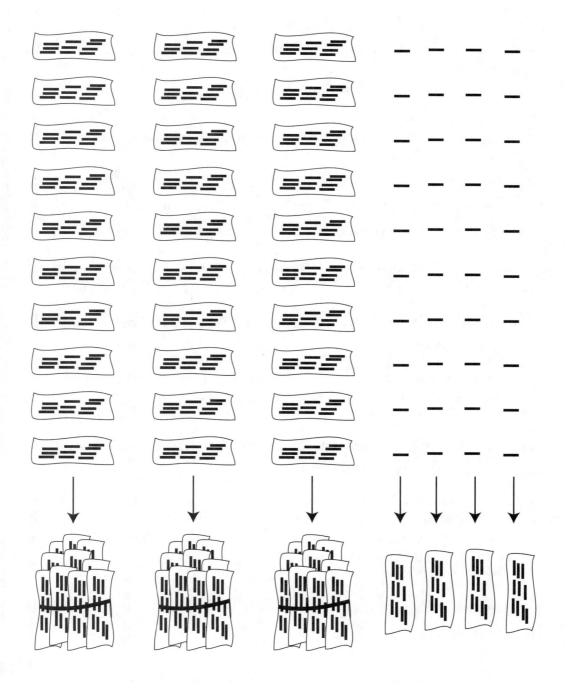

Figure 4.10: 10 Groups of 34 is Bundled into 3 Hundreds and 4 Tens

multiplying by 100 has the same effect as multiplying by 10 twice. Therefore multiplying by 100 moves each digit in a decimal *two places* to the left. Similarly, because

$$1000 = 10 \times 10 \times 10,$$

multiplying by 1000 has the same effect as multiplying by 10 three times. Therefore multiplying by 1000 moves each digit in a decimal *three places* to the left.

Class Activity 4B: If We Wrote Numbers Differently, Multiplying by 10 Might Not be so Easy

Class Activity 4C: Multiplying by Powers of 10 Explains the Cycling of Decimal Representations of Fractions

Exercises for Section 4.2 on Why Multiplying Decimals by 10 is Easy

1. Using the example 10×3.4, explain why we move the digits in a decimal one place to the left when we multiply by 10.

Answers to Exercises for Section 4.2 on Why Multiplying Decimals by 10 is Easy

1. Using bundled toothpicks, the decimal 3.4 can be represented shown in Figure 4.9, as long as 1 toothpick represents $\frac{1}{10}$, and a bundle of 10 toothpicks represents 1. Now use the explanation that is given in the text and that goes along with Figure 4.10.

Problems for Section 4.2 on Why Multiplying Decimals by 10 is Easy

1. Mary says that $10 \times 3.7 = 3.70$. Why might Mary think this? Explain to Mary why her answer is not correct and why the correct answer is right. If you tell Mary a procedure, be sure to tell Mary why the procedure makes sense.

2. (a) Find the decimal representation of $\frac{1}{37}$ to at least 6 places (or as many as your calculator shows). Notice the repeating pattern.

   (b) Now find the decimal representations of $\frac{10}{37}$ and of $\frac{26}{37}$ to at least 6 places. Compare the repeating patterns to each other and to the decimal representation of $\frac{1}{37}$. What do you notice?

   (c) Write $10 \times \frac{1}{37} = \frac{10}{37}$ and $100 \times \frac{1}{37} = \frac{100}{37}$ as mixed numbers (with a whole number part and a fractional part).

   (d) What happens to the decimal representation of a number when it is multiplied by 10, or by 100? Use this, and part (c), to explain the relationships you noticed in part (b).

3. (a) Find the decimal representation of $\frac{1}{41}$ to at least 8 decimal places (preferably 10 or more). Notice the repeating pattern.

   (b) Use your answer in part (a) to find the decimal representations of the numbers

   $$10 \times \frac{1}{41}, \quad 100 \times \frac{1}{41}, \quad 1000 \times \frac{1}{41},$$

   $$10,000 \times \frac{1}{41}, \quad 100,000 \times \frac{1}{41}$$

   without a calculator.

   (c) Write the numbers

   $$10 \times \frac{1}{41} = \frac{1}{41}, \quad 100 \times \frac{1}{41} = \frac{100}{41}, \quad 1000 \times \frac{1}{41} = \frac{1000}{41},$$

   $$10,000 \times \frac{1}{41} = \frac{10,000}{41}, \quad 100,000 \times \frac{1}{41} = \frac{100,000}{41}$$

   as mixed numbers (a whole number part and a fractional part).

   (d) Use your answers in part (b) to find the decimal representations of the fractional parts of the mixed numbers you found in part (c). Do not use your calculator or do long division, use part (b). Explain your reasoning.

4.3   Multiplication and Areas of Rectangles

In this section, we will examine the natural relationship between multiplication and area of rectangles. You are probably familiar with the *length times width*, $L \times W$, formula for areas of rectangles. We will see that *we can explain why this formula is valid by using the meaning of multiplication.*

Area is usually measured in square units. Depending on what we are describing—the page of a book, the floor of a room, a football field, a parcel of land—we usually measure area in square inches, square feet, square yards, square miles, square centimeters, square meters, or square kilometers. A **square inch**, often written 1 in^2, is the area of a square that is 1 inch wide and one inch long, as is the square in Figure 4.11. Similarly, one **square**

Figure 4.11: A 1 Inch by 1 Inch Square Has Area 1 in^2

foot, often written 1 ft^2, is the area of a square that is 1 inch wide and 1 inch long. In general, for any unit of length, one **square unit** is the area of a square that is 1 unit wide and 1 unit long.

The rectangle in Figure 4.12 is three inches wide and two inches long (or high). Why can its area be calculated by multiplying its length times its width? When we subdivide the rectangle into 1 inch × 1 inch squares we have two rows of these squares, and each row has three squares. In other words, the rectangle is made up of 2 groups with 3 squares in each group. Therefore, *according to the meaning of multiplication*, the rectangle is made up of 2 × 3, or 6 squares. Because each 1 inch by 1 inch square has area one square inch, therefore the area of the whole rectangle is 2 × 3 square inches. The same reasoning can be used to determine areas of other rectangles.

In general, if L and W are any whole numbers, then a rectangle that is L units long and W units wide can be subdivided into L rows of 1 unit by 1 unit squares, with W squares in each row, as indicated in Figure 4.13. In other words, the rectangle consists of L groups, with W squares in each

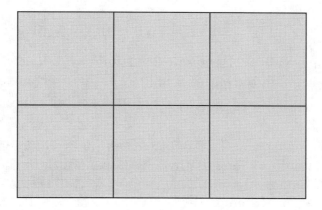

Figure 4.12: A 2 Inch by 3 Inch Rectangle

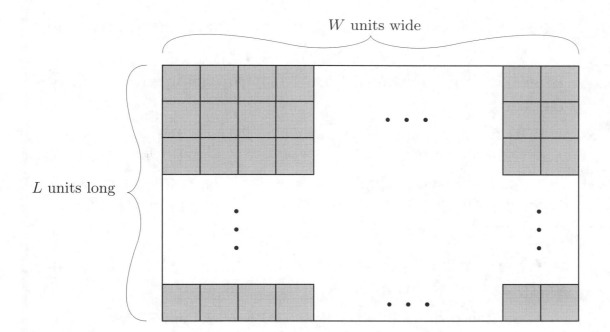

Figure 4.13: A Rectangle that is L Units Long and W Units Wide

group. Therefore, *according to the meaning of multiplication*, the rectangle is made of

$$L \times W$$

1 unit by 1 unit squares. Because each 1 unit by 1 unit square has area 1 square unit, therefore the entire L unit by W unit rectangle has area

$$L \times W \quad \text{square units.}$$

The above explains why the familiar *length times width* formula for areas of rectangles is valid—as long as the length and width of the rectangle are *whole numbers* of units. Later in the chapter we will see that the area formula, and the line of reasoning we used above, is still valid for other positive numbers.

Class Activity 4D: Areas of Rectangles in Square Yards and Square Feet

Class Activity 4E: Using Multiplication to Estimate How Many

Exercises for Section 4.3 on Multiplication and Area of Rectangles

1. Use the *meaning of multiplication* to explain why the area of a carpet that is 20 feet long and 12 feet wide is 20×12 square feet.

2. 1 mile is 1760 yards. Does this mean that 1 square mile is 1760 square yards? If not, how many square yards are in a square mile? Explain carefully.

Answers to Exercises for Section 4.3 on Multiplication and Area of Rectangles

1. You can think of a carpet that is 20 feet long and 12 feet wide as made up of 20 rows of squares with 12 squares in each row, each square being 1 foot wide and 1 foot long. In other words, the carpet consists of 20 groups, with 12 squares in each group. Therefore, according to the meaning of multiplication, the carpet is made of 20×12 squares.

Because the area of a square that is 1 foot wide and 1 foot long is 1 square foot, therefore the area of the carpet is 20×12 square feet.

2. 1 square mile is the area of a square that is 1 mile long and 1 mile wide. Such a square is 1760 yards long and 1760 yards wide, so we can mentally decompose this square into 1760 rows, each having 1760 squares that are 1 yard by 1 yard. Therefore, according to the meaning of multiplication, there are

$$1760 \times 1760 = 3,097,600$$

1 yard by 1 yard squares in a square mile. This means that 1 square mile is $3,097,600$ square yards and not 1760 square yards.

Problems for Section 4.3 on Multiplication and Area of Rectangles

1. 1 foot is 12 inches. Does this mean that 1 square foot is 12 square inches? Draw a picture showing how many square inches are in a square foot. Use the meaning of multiplication to explain why you can calculate the area of 1 square foot in terms of square inches by multiplying.

2. A room has a floor area of 48 square yards. What is the area of the room in square feet? Solve this problem in *two different ways*, each time referring to the meaning of multiplication.

3. 1 kilometer is 1000 meters. Does this mean that 1 square kilometer is 1000 square meters? If not, what is 1 square kilometers in terms of square meters? Explain your answer in detail, referring to the meaning of multiplication.

4. A standard piece of paper is 11 inches long and $8\frac{1}{2}$ inches wide. Use the meaning of multiplication to explain why the area of a standard piece of paper is

$$11 \times 8\frac{1}{2} \text{ in}^2.$$

5. Imagine that you are standing on a sandy beach, like the one in Figure 4.14, gazing off into the distance. How many grains of sand might

you be looking at? To solve this problem, make reaonable assumptions about how wide the beach is and how far down the length of the beach you can see. Explain why your assumptions are reasonable. Based on your assumptions, make a calculation that will give you a reasonably good estimate for the number of grains of sand that you can see.

Figure 4.14: How Many Grains of Sand Do We See Looking Down a Sandy Beach?

6. (a) Explain how Figure 4.15 shows that

$$1 + 2 = \frac{1}{2}(2 \times 3)$$

and

$$1 + 2 + 3 = \frac{1}{2}(3 \times 4).$$

(b) Draw a similar picture for 1+2+3+4 and explain how to generalize the picture to help you calculate other sums, such as $1 + 2 + 3 + \ldots + 10$.

(c) Calculate $1 + 2 + 3 + \ldots + 500$. Have you seen a similar problem before?

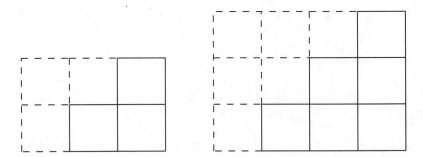

Figure 4.15: Pictures to Help Find Sums

7. The Browns need new carpet for a room with a rectangular floor that is 35 feet wide and 43 feet long. To save money, they will install the carpet themselves. The carpet comes off a large roll that is three *yards* wide. The carpet store will cut off any length of carpet they like, but the Browns must buy the full three yards in width.

   (a) Draw clear, detailed pictures showing two different ways the Browns could lay their carpet.

   (b) For each way of laying the carpet, find how much carpet the Browns will need to buy from the carpet store. Which way will be less expensive for the Browns?

8. The Smiths will be carpeting a room in their house. In one store, they see a carpet they like that costs $35 per square yard. Another store has a similar carpet for $3.95 per square foot. Is this more or less expensive than the carpet at the first store?

9. One acre is 43,560 square feet. If a square piece of land is 3 acres, then what is the length and width of this piece of land in feet? (Remember that the length and width of a square are the same.)

4.4 The Commutative Property of Multiplication

In Chapter 3 we studied some of the properties that the operation of addition enjoys. One of those properties was the commutative property. In this section, we will study the commutative property of *multiplication*. We will explain why this property makes sense for counting numbers by working with areas of rectangles. We will also see ways that we use the commutative property of multiplication.

The **commutative property of multiplication** says that if A and B are any real numbers, then

$$A \times B = B \times A,$$

in other words, $A \times B$ is equal to $B \times A$. For example,

$$297 \times 43 = 43 \times 297,$$

and

$$1.3 \times 5.7 = 5.7 \times 1.3.$$

We assume that this property always holds for *any* pair of numbers, but we can explain why this property makes sense for counting numbers.

If you think of the commutative property of multiplication in terms of groupings, it can seem somewhat mysterious: why do 3 groups of 5 marbles have the same total number of marbles as 5 groups of 3 marbles? Why do 297 baskets of potatoes with 43 potatoes in each basket contain the same total number of potatoes as 43 baskets of potatoes with 297 potatoes in each basket? Of course, we can *calculate* 3×5 and 5×3 and see that they are equal, and we can *calculate* 297×43 and 43×297 and see that they are equal, but by simply calculating, we don't see why the commutative property should hold in other situations as well. A *conceptual* explanation that will show us why this property makes sense will help us understand its meaning better. Such a conceptual explanation is provided by working with arrays or with areas of rectangles.

Suppose you have 3 groups with 5 marbles in each group. According to the meaning of multiplication, this is a total of

$$3 \times 5$$

marbles. You can arrange the 3 groups of marbles so that they form the 3 rows of an array, as pictured on the left in Figure 4.16. But if you now choose the *columns* of the array to be the groups, as seen on the right in Figure 4.16, then there are 5 groups with 3 marbles in each group. Therefore, according to the meaning of multiplication, there is a total of

$$5 \times 3$$

marbles. Since the total number of marbles is the same, either way you count them, therefore

$$3 \times 5 = 5 \times 3.$$

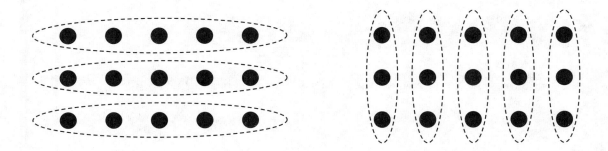

Figure 4.16: $3 \times 5 = 5 \times 3$

By imagining pictures like Figure 4.16, it's easy to see why the commutative property of multiplication should hold for counting numbers other than 3 and 5. Even though it would be ridiculously tedious to draw 297 horizontal rows with 43 potatoes in each row, we can imagine such a picture as similar to Figure 4.16. Such a picture, and the argument above about changing groups from rows to columns, explains why

$$297 \times 43 = 43 \times 297.$$

In these situations there is nothing special about the numbers 3, 5, 297, and 43. So if A and B are *any* counting numbers, there will be a picture similar to Figure 4.16 illustrating that

$$A \times B = B \times A.$$

We can also explain why the commutative property of multiplication makes sense by using areas of rectangles. For example, a rug that is 3 feet by 5 feet can be thought of as made of 3 rows, with 5 squares, each 1 foot by 1 foot, in each row. Therefore the area of the rug is

$$3 \times 5 \text{ square feet.}$$

On the other hand, the rug can be thought of as made of 5 columns, with 3 squares, each 1 foot by 1 foot, in each column, as shown in Figure 4.17. Therefore the area of the rug is

$$5 \times 3 \text{ square feet.}$$

But the area is the same either way you calculate it, therefore

$$3 \times 5 = 5 \times 3.$$

As before, the same line of reasoning will work when other counting numbers replace 3 and 5.

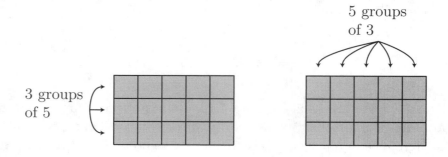

Figure 4.17: $3 \times 5 = 5 \times 3$

Using the Commutative Property of Multiplication

It is easy to use the commutative property of multiplication without even being consciously aware that you are doing so. Suppose you are asked to solve the following problem mentally:

what are fifteen twos?

Rather than counting by twos fifteen times, you might spontaneously turn this problem into:

what are two fifteens?

The new problem is easy to solve mentally by adding 15 and 15 to get 30. When you change the original problem to the new problem, you use the commutative property of multiplication. This is because "fifteen twos" means 15 groups of 2, which is 15×2, and "two fifteens" means 2 groups of 15, which is 2×15. You can substitute the problem 2×15 for the problem 15×2 because, according to the commutative property of multiplication, their results are equal, in other words

$$15 \times 2 = 2 \times 15.$$

Exercises for Section 4.4 on the Commutative Property of Multiplication

1. Use the *meaning of multiplication* and a picture to give a conceptual explanation for why $2 \times 4 = 4 \times 2$. Your explanation should be *general* in the sense that it should be clear why it would remain valid if other counting numbers were to replace 2 and 4.

2. Give an example to show how to use the commutative property of multiplication to make a mental math problem easier to do.

Answers to Exercises for Section 4.4 on the Commutative Property of Multiplication

1. See Figure 4.18. It shows that the total number of objects in 2 groups with 4 objects in each group, which is 2×4 objects, can also be thought of as made out of 4 groups with 2 objects in each group, which is 4×2 objects. Because you have the same number of objects either way you count them, therefore

$$2 \times 4 = 4 \times 2.$$

2. See text.

Problems for Section 4.4 on the Commutative Property of Multiplication

1. Here is Amy's explanation for why the commutative property of multiplication makes sense for counting numbers:

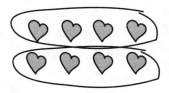   

two groups of 4   four groups of 2

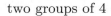

Figure 4.18: $2 \times 4 = 4 \times 2$

Whenever I take two counting numbers and multiply them
I always get the same answer as if I multiply them in the
reverse order. For example,

$$6 \times 8 = 48$$
$$8 \times 6 = 48,$$

$$9 \times 12 = 108$$
$$12 \times 9 = 108$$

$$3 \times 15 = 45$$
$$15 \times 3 = 45$$

It always works that way, no matter which numbers you multiply, you will get the same answer either way you multiply
them.

Did Amy give a convincing explanation for *why* the commutative property of multiplication makes sense for counting numbers? Discuss
Amy's work, referring to some or all of the criteria for a good explanation that are given on page 11.

2. Sue and Tonya started off at the same time with the same job and
   identical salaries. After one year, Sue got a 5% raise and Tonya got
   a 6% raise. The following year, the situation was reversed: Sue got a
   6% raise and Tonya got a 5% raise. After the first year, Tonya's salary
   was higher than Sue's, of course, but whose salary was higher after
   both raises? Solve this problem using the fact that if someone's salary

goes up by 5%, then their new salary is 1.05 times their old salary (similarly for a 6% raise, of course). Using this method, explain how the *commutative property of multiplication* is relevant to comparing the two women's salaries after both raises.

3. A dress is marked down 25% and then it is marked down 20% from the discounted price.

   (a) By what percent is the dress marked down all together (after both discounts)?

   (b) Does the dress cost the same, less or more than if the dress was marked down by 45% from the start? Explain how you can determine the answer to this question without doing any calculating.

   (c) If the dress was marked down by 20% first and then by 25% do you get a different answer to part (a)? Explain how the *commutative property of multiplication* is relevant to this question.

4.5 Multiplication and Volumes of Boxes

Not only are *areas* naturally related to multiplication, but *volumes* are as well, as we'll see in this section. In the next section, we will use volumes to discuss the associative property of multiplication.

Depending on what you want to measure—a dose of liquid medicine, the size of a compost pile, the volume of coal in a mountain—volume can be measured in cubic inches, cubic feet, cubic yards, cubic miles, cubic centimeters, cubic meters or cubic kilometers. (Volumes of liquids are also commonly measured in liters, milliliters, gallons, quarts, cups, and fluid ounces.)

One **cubic centimeter**, often written 1 cm^3, is the volume of a cube that is one centimeter high, one centimeter deep, and one centimeter wide. A drawing of such a cube is shown in Figure 4.19, along with a cube of volume 1 **cubic inch**, 1 inch^3. One **cubic yard**, written 1 yd^3, is the volume of a cube that is one yard high, one yard deep and one yard wide. In general, for any unit of length, one **cubic unit**, or 1 unit^3, is the volume of a cube that is 1 unit high, 1 unit deep, and 1 unit wide.

We will be working with volumes of boxes and box-shapes (these shapes are also called **rectangular prisms**). The volume, in cubic units, of a box or box-shape is just the number of 1 unit by 1 unit by 1 unit cubes that it would

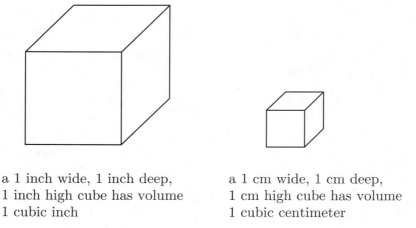

a 1 inch wide, 1 inch deep,   a 1 cm wide, 1 cm deep,
1 inch high cube has volume   1 cm high cube has volume
1 cubic inch   1 cubic centimeter

Figure 4.19: Cubes of Volume 1 inch^3 and 1 cm^3

take to fill the box or make the box-shape. You may remember a formula for
the volume of a box, but if so, pretend for a moment that you don't know
this formula. We will see how to derive the formula for the volume of a box
from the *meaning of multiplication*.

Suppose you have a box that is 4 inches high, 2 inches deep, and 3 inches
wide, as pictured in Figure 4.20). What is the volume of this box? If you

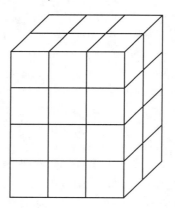

Figure 4.20: A 4 Inch High, 2 Inch Deep, 3 Inch Wide Box

have a set of building blocks that are all one inch high, one inch deep, and
one inch wide, then you can use the building blocks to build a box-shape that

would fill the box. The number of blocks needed is the volume of the box in cubic inches. We can use multiplication to describe the number of blocks needed by considering the box as made of 4 horizontal layers, as shown in Figure 4.21. Each layer consists of two rows of three blocks, so *according to*

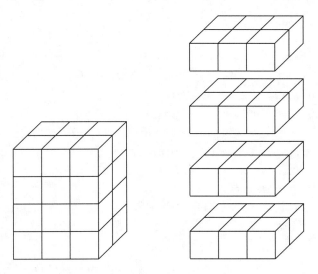

Figure 4.21: A box divided into layers

the meaning of multiplication, each layer contains 2×3 blocks. There are 4 layers, each containing 2×3 blocks, so once again, *according to the meaning of multiplication*, there are

$$4 \times (2 \times 3)$$

blocks making up the box shape. Therefore the box has a volume of

$$4 \times (2 \times 3) = 4 \times 6 = 24 \text{ cubic inches.}$$

The reasoning used above applies generally. Suppose you have a box that is H units high, D units deep and W units wide. If H, D and W are whole numbers, then as before, you can build such a box-shape out of 1 unit by 1 unit by 1 unit cubes. How many cubes does it take? If you consider the box as made of horizontal layers, then there are H layers. Each layer is made up of D rows of W blocks (or you can think of it as W rows of D blocks) and therefore contains $D \times W$ blocks, according to the meaning of multiplication. Because there are H layers with $D \times W$ blocks in each layer, therefore, according to the meaning of multiplication, the box is made out of

$$H \times (D \times W)$$

1 unit by 1 unit by 1 unit cubes. Therefore the box has volume

$$H \times (D \times W) \text{ cubic units.}$$

Notice that the order in which the letters H, D, and W appear and the way the parentheses are placed in the expression $H \times (D \times W)$ corresponds to the way the box was divided into groups.

As with areas, it turns out that this *height times depth times width* formula for the volume of a box remains valid even when the height, depth, and width of the box are not a whole numbers, and the same reasoning as above applies. The volume of a box that is H units high, D units deep, and W units wide is always $H \times (D \times W)$ cubic units.

In the next section we will examine the use of parentheses in expressions like $H \times (D \times W)$.

Class Activity 4F: Ways to Describe the Volume of a Box with Multiplication

Class Activity 4G: Volumes of Boxes in Cubic Yards and Cubic Feet

Class Activity 4H: How Many Gumdrops?

Exercises for Section 4.5 on Multiplication and Volumes of Boxes

1. Use the *meaning of multiplication* to explain why a box that is 3 feet wide, 2 feet deep, and 4 feet high has volume $4 \times (2 \times 3)$ cubic feet.

2. How many cubic feet of mulch will you need to cover a garden that is 10 feet wide and 4 yards long with 2 inches of mulch? Can you find the amount of mulch you need by multiplying $2 \times (10 \times 4)$?

3. One cubic foot of water weighs about 62 pounds. How much will the water in a swimming pool that is 20 feet wide, 30 feet long, and 4 feet deep weigh?

Answers to Exercises for Section 4.5 on Multiplication and Volumes of Boxes

1. See the explanation in the text. Substitute "feet" where the text says "inches."

2. Since 1 yard = 3 feet, and 1 foot = 12 inches, therefore 4 yards = 12 feet, and 2 inches = $\frac{1}{6}$ feet. So you can think of the mulch as forming a box shape that is $\frac{1}{6}$ feet high, 10 feet deep (or wide) and 12 feet wide (or deep). This box has volume

$$\frac{1}{6} \times (10 \times 12) \text{ cubic feet } = 20 \text{ cubic feet },$$

so you'll need 20 cubic feet of mulch.

No, you can't find the amount of mulch you need by multiplying $2 \times 10 \times 4$ because each number refers to a different unit of length.

3. Because the water in the pool is in the shape of a box that is 4 feet high, 20 feet wide, and 30 feet deep, it can be thought of as made of

$$4 \times (20 \times 30) = 2400$$

1 foot by 1 foot by 1 foot cubes of water. Each of those cubes of water weighs 62 pounds, so the water in the pool will weigh

$$2400 \times 62 \text{ pounds } = 148,800 \text{ pounds}.$$

Problems for Section 4.5 on Multiplication and Volumes of Boxes

1. One foot is 12 inches. Does this mean that one cubic foot is 12 cubic inches? Describe how to use the meaning of multiplication to determine what 1 cubic foot is in terms of cubic inches.

2. You will need 60 pennies or other small, stackable objects (such as blocks or snap-together cubes) to do this problem. Arrange 60 pennies into 12 stacks with 5 pennies in each stack. Arrange the 12 stacks of pennies into a rectangular shape consisting of 3 rows with 4 stacks in each row. *With the pennies arranged in this fashion* describe 6 different

ways of breaking the pennies into natural groups with the same number of pennies in each group. In each case, give a corresponding expression for the total number of pennies based on the *meaning of multiplication* as given in the text.

3. Figure 4.22 shows a grocery store display of cases of sodas. The display consists of a large box shape with a "staircase" on top. The display is 7 cases wide and 5 cases deep; it is 4 cases tall in the front, and 8 cases tall at the back. How many cases of sodas are there in the display? Solve this problem in *at least two different ways* and explain your method in each case.

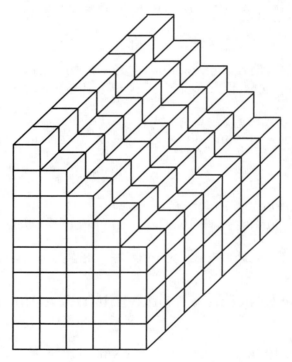

Figure 4.22: A Display Of Cases Of Soda

4. Investigate the following two questions and explain your conclusions.

   (a) If you make a rectangular garden that is twice as wide and twice as long as a rectangular garden that you already have, how will the area of the larger garden compare to the area of the original garden? (Will the larger garden be twice as big, 3 times as big, 4 times as big etc?)

(b) If you make a cardboard box that is twice as wide, twice as tall and twice as deep as a cardboard box that you already have, how will the volume of the larger box compare to the volume of the original box? (Will the larger box be twice as big, 3 times as big, 4 times as big etc?)

5. Give a good estimate for how many neatly stacked hundred-dollar bills you could fit in a briefcase that is 20 inches long, 11 inches wide, and 2 inches thick on the inside. Describe your method and explain why it gives a good estimate.

6. A cube that is 10 inches wide, 10 inches deep and 10 inches high is made out of smaller cubes that are each 1 inch wide, 1 inch deep and 1 inch high. The large cube is then painted on the outside. How many of the smaller cubes (that make up the large cube) have paint on them?

7. The Better Baking Company is introducing a new line of reduced fat brownies in addition to their regular brownies. The batter for the reduced fat brownies contains $\frac{1}{3}$ less fat than the batter for the regular brownies. Both types of brownies will be baked in the same size rectangular pan which is 24 inches wide and 30 inches long. The regular brownies are cut from this pan by dividing the width into 12 equal segments and by dividing the length into 10 equal segments (so that each regular brownie is 2 inches by 3 inches). Each regular brownie contains 6.3 grams of fat. In addition to using a reduced fat batter, the Better Baking Company would like to further reduce the amount of fat in their new brownies by making these brownies smaller than the regular ones. You have been contacted to help with this task. Present two different ways that the length and width of the pan could be divided to produce smaller brownies. In each case, calculate the amount of fat in each brownie and explain the basis for your calculation. The brownies should be of a reasonable size and there should be no waste left over in the pan after cutting the brownies. The length and width of the brownies do not necessarily have to be whole numbers of inches.

4.6   The Associative Property of Multiplication

The **associative property of multiplication** says that if A, B, and C are any real numbers, then

$$(A \times B) \times C = A \times (B \times C).$$

We assume that this property holds for all real numbers, but as we'll see in this section, we can explain why it makes sense for counting numbers. We will also see how to use the associative property of multiplication.

Explaining Why the Associative Property of Multiplication Makes Sense

Why is the associative property valid? For example, why is

$$(4 \times 2) \times 3 = 4 \times (2 \times 3)?$$

In other words,

> why is the quantity (4×2) times 3 equal to 4 times the quantity (2×3)?

In this example, because specific numbers are involved, we can evaluate the quantities on both sides of the $=$ sign and see that they really are equal:

$$(4 \times 2) \times 3 = 8 \times 3 = 24$$

and

$$4 \times (2 \times 3) = 4 \times 6 = 24,$$

therefore

$$(4 \times 2) \times 3 = 4 \times (2 \times 3).$$

But when we evaluate the expressions, we only show that the associative property holds *in this one special case*. Why does it make sense that the associative property of multiplication holds for *all* real numbers?

To explain why the associative property of multiplication makes sense for all counting numbers, we will develop a *conceptual* explanation for why

$$(4 \times 2) \times 3 = 4 \times (2 \times 3).$$

This explanation will be *general* in the sense that it will also explain why the equation is true if we were to replace the numbers 4, 2, and 3 with other counting numbers. Our explanation is based on calculating the volume of a box in two different ways.

In the previous section, we decomposed a 4 inch high, 2 inch deep, 3 inch wide box-shape into 4 groups of blocks, with 2×3 blocks in each group, as shown on the left of Figure 4.23. Viewed that way, the total number of blocks

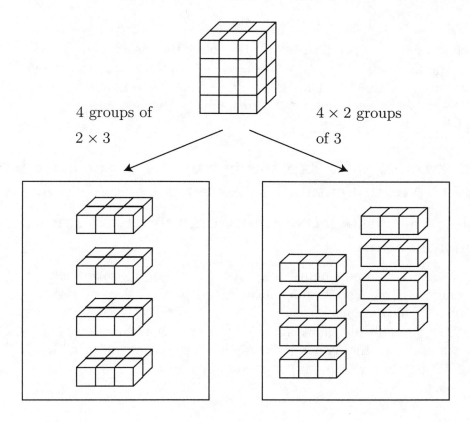

Figure 4.23: Showing that $4 \times (2 \times 3) = (4 \times 2) \times 3$

in the box-shape is

$$4 \times (2 \times 3).$$

On the other hand, the box-shape can be decomposed into 4×2 groups, with 3 blocks in each group, as shown on the right of Figure 4.23. Therefore, according to the meaning of multiplication, there are

$$(4 \times 2) \times 3$$

blocks in the box-shape. There are 4×2 groups because the groups of 3 blocks can themselves be arranged into 4 groups of 2. Because the total number of blocks in the box-shape is the same either way we count them, therefore

$$(4 \times 2) \times 3 = 4 \times (2 \times 3),$$

which is what we wanted to establish. Notice that the same argument works when the numbers 4, 2, and 3 are replaced with other counting numbers— only the size of the box will change. Therefore the reasoning above explains why the associative property of multiplication makes sense for all counting numbers.

Class Activity 4I: Explaining why the Associative Property of Multiplication Makes Sense

Using the Associative and Commutative Properties of Multiplication

The associative and commutative properties of multiplication allow for great flexibility in carrying out multiplication problems. They are especially useful for mental calculations.

When we multiply many numbers together, it is common not to use any parentheses, and the associative property of multiplication allows us to do this. Rather than writing an expression like

$$(17 \times 25) \times 4$$

or

$$17 \times (25 \times 4)$$

we can simply write the expression

$$17 \times 25 \times 4$$

without any parentheses, and we are free to multiply adjacent numbers in whatever order we choose. In this case, it's easiest to first multiply the 25 with the 4 to get 100, and then multiply 17 with 100 to get 1700. The following sequence of equations correspond to this strategy:

$$
\begin{aligned}
17 \times 25 \times 4 &= 17 \times (25 \times 4) \\
&= 17 \times 100 \\
&= 1700.
\end{aligned}
$$

Rather than writing an expression like

$$(41 \times 2) \times 3) \times 5,$$

according to the associative property of multiplication we can simply write

$$41 \times 2 \times 3 \times 5,$$

and we are free to multiply adjacent numbers in any order we please. Furthermore, we can invoke the commutative property of multiplication and switch the order in which numbers appear. In this case, it would be convenient to multiply the 2 with the 5, so we can switch the placement of the 3 and the 5 so that the 2 is adjacent to the 5:

$$41 \times 2 \times 5 \times 3.$$

The following equations correspond to this strategy, and show how to complete the calculation:

$$
\begin{aligned}
41 \times 2 \times 3 \times 5 &= 41 \times 2 \times 5 \times 3 \\
&= 41 \times 10 \times 3 \\
&= 410 \times 3 \\
&= 1230.
\end{aligned}
$$

By using the associative and commutative property of multiplication, we can often make mental multiplication problems easier to solve.

Class Activity 4J: Using the Associative Property of Multiplication

Exercises for Section 4.6 on the Associative Property of Multiplication

1. Explain how you use the associative property of multiplication when you calculate 7×600 mentally.

2. Use the meaning of multiplication and the idea of decomposing a box-shape in two different ways to explain why $4 \times (2 \times 3) = (4 \times 2) \times 3$.

3. Describe how to use the design in Figure 4.24 to give a general, conceptual explanation for why $5 \times (2 \times 2) = (5 \times 2) \times 2$.

one spiral

Figure 4.24: A Design Of Spirals

Answers to Exercises for Section 4.6 on the Associative Property of Multiplication

1. When we calculate 7×600 mentally, we first calculate 7×6 (a "basic fact") and then we multiply by 100. This means that the 6 in 600 goes from being grouped with 100 (to make 600) to being grouped with 7. In equation form we can express this as follows:

$$
\begin{aligned}
7 \times 600 &= 7 \times (6 \times 100) \\
&= (7 \times 6) \times 100 \\
&= 42 \times 100 \\
&= 4200.
\end{aligned}
$$

The associative property of multiplication is used at the second equal sign.

2. See text.

3. See Figure 4.25. On the one hand, the design can be thought of as made up of 5 clusters of spirals, with 2 groups of 2 spirals in each

cluster. This means there are $5 \times (2 \times 2)$ spirals in the design. On the other hand, the design can be thought of as made up of 5×2 groups of spirals with 2 spirals in each group. There are 5×2 groups because there are 5 sets of 2 groups. This means there are $(5 \times 2) \times 2$ spirals in the design. Because there are the same number of spirals, no matter how they are counted, therefore

$$5 \times (2 \times 2) = (5 \times 2) \times 2.$$

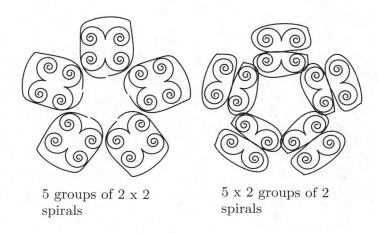

5 groups of 2 x 2
spirals

5 x 2 groups of 2
spirals

Figure 4.25: $5 \times (2 \times 2) = (5 \times 2) \times 2$

Problems for Section 4.6 on the Associative Property of Multiplication

1. Write equations to show how the commutative and associative properties of multiplication are involved when you calculate 40×800 mentally by relying on "basic multiplication facts" (such as 4×8). Write your equations in the form

$$
\begin{aligned}
40 \times 800 &= \text{some expression} \\
&= \vdots \\
&= \text{some expression.}
\end{aligned}
$$

Indicate specifically where the commutative and associative properties of multiplication are used.

2. Explain how to use the associative property of multiplication to make 16×25 easy to calculate mentally. Write equations that show why your method is valid and show specifically where you have used the associative property of multiplication. Write your equations in the form

$$
\begin{aligned}
16 \times 25 &= \quad \text{some expression} \\
&= \quad \vdots \\
&= \quad \text{some expression.}
\end{aligned}
$$

3. Use the facts that

$$
\begin{aligned}
1 \text{ mile} &= 1760 \text{ yards} \\
1 \text{ yard} &= 3 \text{ feet} \\
1 \text{ foot} &= 12 \text{ inches}
\end{aligned}
$$

in order to calculate the number of inches in a mile. *Do this in two different ways, so as to illustrate the associative property of multiplication.*

4. Julia says that it's easy to multiply a number by 4 because you just "double the double". Explain Julia's idea and explain why it uses the associative property of multiplication.

5. Carmen says that it's easy to multiply an even number by 5 because you just take half of the number. Elaborate on Carmen's idea, giving a precise statement about multiplying even whole numbers by 5. Write equations and use the *associative property of multiplication* to explain why your precise statement is true. (Remember that an even number is an integer that is 2 times another integer. So for example, 74 is even because $74 = 2 \times 37$.)

4.7　The Distributive Property

In the previous chapter, we studied the associative and commutative properties of addition. So far in this chapter we have studied the associative and commutative properties of multiplication. In this section we will study the only property of arithmetic that involves both addition and multiplication, namely the distributive property of multiplication over addition. The distributive property is the most important and computationally powerful tool in all of arithmetic. It allows for tremendous flexibility in performing mental calculations, and, as we will see in the next section, the distributive property underlies the standard longhand multiplication technique.

Before we discuss the distributive property of multiplication over addition, we will discuss the conventions for interpreting expressions involving both multiplication and addition.

Expressions Involving Both Multiplication and Addition

Expressions that involve both addition and multiplication must be interpreted suitably, according to the conventions developed by mathematicians. This situation is entirely unlike the situation where *only* addition, or *only* multiplication is involved in an expression. In those situations, parentheses can safely be dropped and adjacent numbers can be combined at will. However, *when both multiplication and addition are present in an expression, parentheses cannot generally be dropped* without changing the value of the expression. For example,

$$7 + 5 \times 2$$

is *not equal* to

$$(7 + 5) \times 2.$$

To properly interpret an expression such as

$$7 + 5 \times 2$$

or

$$5 \times 17 + 9 \times 10^2 - 12 \times 94 + 20 \div 5$$

we need to use the following conventions:

all powers are calculated first,

next *multiplications and divisions are performed* from left to right,

finally *additions and subtractions are performed.*

Expressions inside parentheses are always evaluated first, subject to the conventions above.

Therefore

$$7 + 5 \times 2 \; = \; 7 + 10$$
$$= \; 17,$$

whereas

$$(7 + 5) \times 2 \; = \; 12 \times 2$$
$$= \; 24.$$

Similarly,

$$5 \times 17 + 9 \times 10^2 - 12 \times 94 + 20 \div 5$$
$$= \; 5 \times 17 + 9 \times 100 - 12 \times 94 + 20 \div 5$$
$$= \; 85 + 900 - 1128 + 4$$
$$= \; -139.$$

Explaining Why the Distributive Property Makes Sense

The **distributive property of multiplication over addition** says that for all real numbers, A, B and C,

$$A \times (B + C) = A \times B \; + \; A \times C.$$

Notice the use of parentheses to group the B and C: the expression

$$A \times (B + C)$$

means A times the *quantity* $B + C$.

As with the other properties of arithmetic that we have studied, we assume that the distributive property holds for all real numbers. However, we can explain why the distributive property makes sense for counting numbers by decomposing arrays of objects or by decomposing rectangles.

Consider an array of dots consisting of 4 horizontal rows, with 7 dots in each row. The vertical line in the diagram below shows one way to decompose this array of dots into two smaller arrays.

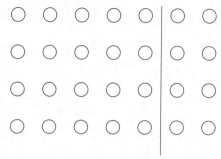

There are two ways of expressing the total number of dots in the array that correspond to this way of decomposing the array. On the one hand, there are 4 rows with $5 + 2$ dots in each row, and therefore the total number of dots is

$$4 \times (5 + 2).$$

(Notice that parentheses are needed to say 4 times the *quantity* 5 plus 2.) On the other hand, the vertical line decomposes the array of dots into two arrays. The array on the left consists of 4 rows with 5 dots in each row, and thus contains of 4×5 dots; the array on the right consists of 4 rows with 2 dots in each row, and therefore contains 4×2 dots. Therefore, combining these two arrays of dots, there is a total of

$$4 \times 5 \; + \; 4 \times 2$$

dots. Because the total number of dots is the same, either way you count them, therefore

$$4 \times (5 + 2) = 4 \times 5 \; + \; 4 \times 2.$$

This explains why the distributive property makes sense in this case. But notice that the reasoning is *general* in the sense that if we were to replace the numbers 4, 5, and 2 with other counting numbers, and if we were to correspondingly adjust the the array of dots, then our argument would still hold. Therefore the distributive property makes sense for all counting numbers.

Class Activity 4K: Using Areas of Rectangles to Explain Why the Distributive Property Makes Sense

Variations on the Distributive Property

Several useful variations on the distributive property are listed below. All these variations can be obtained from the original distributive property by using the commutative property of multiplication, by using the distributive property repeatedly, or by using the fact that

$$B - C = B + (-C).$$

The Original Distributive Property

$$A \times (B + C) = A \times B \ + \ A \times C$$

for all real numbers A, B, and C.

Variation 1

$$(A + B) \times C = A \times C \ + \ B \times C$$

for all real numbers A, B, and C.

Variation 2

$$A \times (B + C + D) = A \times B \ + \ A \times C \ + \ A \times D$$

for all real numbers A, B, C, and D.

Variation 3

$$A \times (B - C) = A \times B \ - \ A \times C$$

for all real numbers A, B, and C.

Class Activity 4L: Variations on the Distributive Property

Using the Distributive Property

Just like the other properties of arithmetic that we have studied, we can often use the distributive property to make mental arithmetic problems easier to solve.

For example, what is an easy way to calculate

$$41 \times 25$$

mentally? We can use the following strategy:

40 times 25 is 1000, plus one more 25 is 1025.

The following sequence of equations correspond to this strategy and show that it uses the first variation on the distributive property (at the second = sign).

$$
\begin{aligned}
41 \times 25 &= (40 + 1) \times 25 \\
&= 40 \times 25 + 1 \times 25 \\
&= 1000 + 25 \\
&= 1025
\end{aligned}
$$

Vignette: How a Fourth Grader Used the Distributive Property to Calculate with Fractions

The distributive property holds for all real numbers, therefore it can be used with fractions. In the following conversation, a fourth grader skillfully uses the distributive property to calculate with fractions. The exchange is taken from *Developing Children's Understanding of the Rational Numbers: A New Model and an Experimental Curriculum* by Joan Moss and Robbie Case [32, page 135]. *"Experimental S1"* is one of the fourth grade students who participated in an experimental curriculum described in the article.

> *Experimenter:* Another student told me that 7 is 3/4 of 10. Is it?
>
> *Experimental S1:* No, because of one half of 10 is 5. One half of 5 is 2 and 1/2. So if you add 2 1/2 to 5, that would be 7 1/2. So 7 1/2 is 3/4 of 10, not 7.

Why is the student's reasoning valid? Notice that the student adds half of 10 and half of half of ten to get 7 and a half. In other words, the student says that 3/4 of 10 is half of ten plus half of half of ten, or:

$$\frac{3}{4} \times 10 = \frac{1}{2} \times 10 + \frac{1}{2} \times \frac{1}{2} \times 10.$$

We can use the following two facts to explain why the student's method is correct:

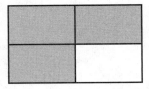

3/4 is half plus half of half

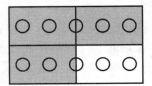

3/4 of 10 is half of 10 plus half of half of 10

Figure 4.26: Finding 3/4 of 10

- Three fourths is a half plus half of a half, or in other words:

$$\frac{3}{4} = \frac{1}{2} + \frac{1}{2} \times \frac{1}{2},$$

as seen in Figure 4.26.

- The distributive property.

Observe how the following equations use these two facts, and show why the student's reasoning is valid:

$$
\begin{aligned}
\frac{3}{4} \times 10 &= \left(\frac{1}{2} + \frac{1}{2} \times \frac{1}{2}\right) \times 10 \\
&= \frac{1}{2} \times 10 + \frac{1}{2} \times \frac{1}{2} \times 10 \\
&= 5 + 2\frac{1}{2} \\
&= 7\frac{1}{2}
\end{aligned}
$$

The distributive property is used at the second = sign.

Class Activity 4M: Solving Arithmetic Problems Mentally

Class Activity 4N: Which Properties of Arithmetic do These Calculations Use?

Class Activity 4O: Writing Equations that Correspond to a Method of Calculation

Class Activity 4P: Mental Math

Class Activity 4Q: Using Properties of Arithmetic

Class Activity 4R: Useful Patterns in the One-Digit Multiplication Tables

Class Activity 4S: The Distributive Property and FOIL

Equations and Properties of Arithmetic are Stepping Stones to Algebra

This section and several others in the book have emphasized the writing of equations and the use of properties of arithmetic to solve arithmetic problems. In some cases, we used properties of arithmetic to help us solve a problem mentally, but in other cases, especially when we write long strings of equations that go along with a method of calculation, it may seem that we are only making the problem longer and harder. So why bother, especially when there are efficient algorithms for carrying out the calculations? The writing of equations to go along with a method of calculation, and the use of properties of arithmetic to calculate are emphasized here *because these are fundamental skills in algebra.* For example, one standard type of problem in algebra is to simplify an expression such as

$$9x^2 + 4x(7 - x).$$

We can solve this problem by writing the following string of equations:

$$
\begin{aligned}
9x^2 + 4x(7 - x) &= 9x^2 + 4x \cdot 7 - 4x \cdot x \\
&= 9x^2 + 28x - 4x^2 \\
&= 9x^2 - 4x^2 + 28 \\
&= 5x^2 + 28.
\end{aligned}
$$

In doing this, we used the distributive property as well as the associative and commutative properties of multiplication and addition.

The National Council of Teachers of Mathematics advocates that *all children learn algebra* (see [40, p. 37]). This ambitious goal can only succeed if children develop a strong foundation to support the learning of algebra. Research shows that the leap from arithmetic to algebra is a difficult one for students because algebra involves dealing with unknown quantities (see [21]). However, research also shows that some of the difficulties in students' algebra learning can already be found when students work with equations in arithmetic (see [28]). It therefore makes sense that elementary school children should learn to calculate flexibly by using properties of arithmetic (even if they do not know the formal names of these properties and are not aware that they are using properties of arithmetic) and should learn to work with equations.

Developing the ability to calculate flexibly by using properties of arithmetic lays a foundation for algebra. But according to the National Council of Teachers of Mathematics, [40, p. 35], there is another benefit:

> Researchers and experienced teachers alike have found that when children in the elementary grades are encouraged to develop, record, explain, and critique on another's strategies for solving computational problems, a number of important kinds of learning can occur...

Such experiences can help children strengthen their understanding of place value and develop better number sense (see [25]).

Curriculum frameworks or guidelines of various states recognize the importance of using properties of arithmetic by recommending or mandating that children learn to use these properties. The Georgia Quality Core Curriculum (QCC) [19] expects that a third grader:

> Uses properties of addition and multiplication (including commutative, associative, and properties of zero and one).

The Mathematics Framework for California Public Schools [9] identifies the following as a key standard for grade 2:

> Use the commutative and associative rules to simplify mental calculations and to check results.

Exercises for Section 4.7 on the Distributive Property

1. Compute mentally $97346 \times 142349 + 2654 \times 142349$.

2. Write equations to go along with the following reasoning for determining 5×7.

   > I know 2×7 is 14. Then another 2×7 makes 28. And one more 7 makes 35.

   Which properties of arithmetic are used? Draw a picture to go along with the reasoning.

3. Use graph paper to draw a picture that shows why

   $$20 \times 19 = 20 \times 20 - 20 \times 1.$$

   Also, use this equation to help you calculate 20×19 mentally.

4. Draw a picture to show that

   $$(10 + 2) \times (10 + 3) = 10 \times 10 + 10 \times 3 + 2 \times 10 + 2 \times 3.$$

   Then write equations that use properties of arithmetic to show why the above equation is true.

5. The string of equations below corresponds to a mental method for calculating 45×11. Explain in words why the method of calculation makes sense. Which properties of arithmetic are used, and where?

   $$
   \begin{aligned}
   45 \times 11 &= 45 \times (10 + 1) \\
   &= 45 \times 10 + 45 \times 1 \\
   &= 450 + 45 \\
   &= 495
   \end{aligned}
   $$

6. Each arithmetic problem below has a description for how to solve the problem. In each case, write a string of equations that correspond to the given description. Identify properties of arithmetic that are used. Write your equations in the following form:

   $$
   \begin{aligned}
   \text{original} &= \text{some expression} \\
   &= \vdots \\
   &= \text{some expression}
   \end{aligned}
   $$

(a) Problem: What is 6×40?

Solution: 6 times 4 is 24, then you multiply by 10 and get 240.

(b) Problem: What is 110% of 62?

Solution: 100% of 62 is 62 and 10% of 62 is 6.2, so all together it's 68.2.

(c) Problem: Calculate the 7% tax on a purchase of $25.

Solution: 10% of $25 is $2.50, so 5% is $1.25. One percent of $25 is 25 cents, so 2% is 50 cents. This means 7% is $1.25 plus 50 cents, which is $1.75.

(d) Problem: What is 45% of 300?

Solution: 50% of 300 is 150. Ten percent of 300 is 30 and half of that is 15, so the answer is 135.

(e) Problem: Find $\frac{3}{4}$ of 72.

Solution: Half of 72 is 36. Half of 36 is 18. Then to add 36 and 18 I did 40 plus 14, which is 54.

7. For each of the following problems, use the distributive property to help make the problem easy to do mentally. In each case, write a string of equations that correspond to your strategies. Write your equations in the following form:

$$\begin{aligned} \text{original} &= \text{some expression} \\ &= \vdots \\ &= \text{some expression} \end{aligned}$$

(a) Calculate 51% of 140.

(b) Calculate 95% of 60.

(c) Calculate $\frac{5}{8} \times 280$.

(d) Calculate 99% of 80.

8. For each of the following arithmetic problems, use properties of arithmetic to make the problem easy to solve the problem mentally. Write a string of equations that corresponds to your method. Say which properties of arithmetic you use and where you use them.

(a) 25×84

(b) 49×6

(c) 486×5

9. Joey, a second grader, used the following method to mentally calculate the number of minutes in a day. First, Joey calculated 25 times 6 by finding 4 times 25 plus two times 25, which is 150. Then he subtracted 6 from 150 to get 144. Then he multiplied 144 by 10 to get the answer: 1440. Write a sequence of equations that correspond to Joey's method and that show why his method is legitimate. What properties of arithmetic are involved?

Answers to Exercises For Section 4.7 on the Distributive Property

1. By the distributive property,

$$
\begin{aligned}
97,346 \times 142,349 + 2,654 \times 142,349 &= (97,346 + 2,654) \times 142,349 \\
&= 100,000 \times 142,349 \\
&= 14,234,900,000.
\end{aligned}
$$

2. The distributive property is used at the second equal sign below.

$$
\begin{aligned}
5 \times 7 &= (2 + 2 + 1) \times 7 \\
&= 2 \times 7 + 2 \times 7 + 1 \times 7 \\
&= 14 + 14 + 7 \\
&= 28 + 7 = 35
\end{aligned}
$$

The following picture corresponds to the arithmetic because it shows 2 sevens, another 2 sevens, and then a single seven, to make a total of 5 sevens.

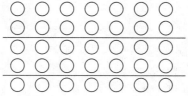

3. Figure 4.27 shows 20 rows of 20 squares. If you take away 20 rows of 1 square—namely a vertical strip of squares—then you are left with 20 rows of 19 squares, thus illustrating that

$$20 \times 19 = 20 \times 20 - 20 \times 1.$$

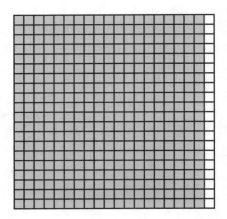

Figure 4.27: $20 \times 19 = 20 \times 20 - 20 \times 1$

4. The picture in Figure 4.28 shows $10 + 2$ groups of $10 + 3$ small squares. These are broken into four groups: 10 groups of 10, 10 groups of 3, 2 groups of 10, and 2 groups of 3. Since it's the same number of small squares no matter how you count them, therefore according to the meaning of multiplication,

$$(10 + 2) \times (10 + 3) = 10 \times 10 + 10 \times 3 + 2 \times 10 + 2 \times 3.$$

Using properties of arithmetic:

$$\begin{aligned}
(10 + 2) \times (10 + 3) &= (10 + 2) \times 10 + (10 + 2) \times 3 \\
&= 10 \times 10 + 2 \times 10 + 10 \times 3 + 2 \times 3
\end{aligned}$$

The distributive property was used at both equal signs. In fact, was used twice at the second equal sign.

10 groups of 10

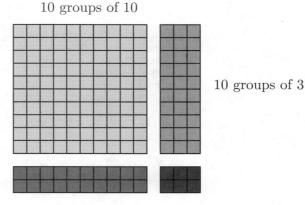

10 groups of 3

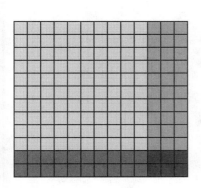

10+2 groups of 10+3

2 groups of 10   2 groups of 3

Figure 4.28: $(10 + 2) \times (10 + 3) = 10 \times 10 + 10 \times 3 + 2 \times 10 + 2 \times 3$

5. 45×11 stands for the total number of objects in 45 groups that have 11 objects in each group. These objects can be broken into 45 groups of 10 and another 45 groups of 1. The number of objects in 45 groups of 10 is 450, and the number of objects in 45 groups of 1 is 45. In all that makes $450 + 45$ which is 495 objects. The distributive property is used to say that $45(10 + 1) = 45 \times 10 + 45 \times 1$.

6. (a)

$$
\begin{aligned}
6 \times 40 &= 6 \times (4 \times 10) \\
&= (6 \times 4) \times 10 \\
&= 24 \times 10 \\
&= 240.
\end{aligned}
$$

The associative property of multiplication is used at the second equal sign in order to switch the placement of the parentheses and group the 4 with the 6 instead of with the 10.

(b)

$$
\begin{aligned}
110\% \times 62 &= (100\% + 10\%) \times 62 \\
&= 100\% \times 62 + 10\% \times 62 \\
&= 62 + 6.2
\end{aligned}
$$

$$= \quad 68.2$$

The distributive property is used at the second equal sign.

(c)

$$
\begin{aligned}
7\% \times 25 \quad &= \quad (5\% + 2\%) \times 25 \\
&= \quad 5\% \times 25 + 2\% \times 25 \\
&= \quad \frac{1}{2} \times 10\% \times 25 + 2 \times 1\% \times 25 \\
&= \quad \frac{1}{2} \times 2.5 + 2 \times .25 \\
&= \quad 1.25 + .50 \\
&= \quad 1.75
\end{aligned}
$$

The distributive property is used at the second equal sign. (Although it is hidden, technically speaking, the associative property of multiplication is actually used twice at the fourth equal sign in order to calculate 10% of 25 and 1% of 25 first, and then multiply those by $\frac{1}{2}$ and 2 respectively.)

(d)

$$
\begin{aligned}
45\% \times 300 \quad &= \quad (50\% - 5\%) \times 300 \\
&= \quad 50\% \times 300 - 5\% \times 300 \\
&= \quad 150 - \frac{1}{2} \times 10\% \times 300 \\
&= \quad 150 - \frac{1}{2} \times 30 \\
&= \quad 150 - 15 \\
&= \quad 135
\end{aligned}
$$

The distributive property is used at the second equal sign. (Although it is hidden, technically speaking, the associative property of multiplication is used at the fourth equal sign in order to calculate 10% of 300 first, and then find half of that.)

(e)

$$\frac{3}{4} \times 72 \quad = \quad (\frac{1}{2} + \frac{1}{2} \times \frac{1}{2}) \times 72$$

$$= \frac{1}{2} \times 72 + \frac{1}{2} \times \frac{1}{2} \times 72$$
$$= 36 + \frac{1}{2} \times 36$$
$$= 36 + 18$$
$$= 36 + (4 + 14)$$
$$= (36 + 4) + 14$$
$$= 40 + 14$$
$$= 54.$$

The distributive property is used at the second equal sign. The associative property of addition was used at the sixth equal sign in order to change the placement of the parentheses. (Although it is hidden, technically speaking, the associative property of multiplication is used at the third equal sign, in order to find half of 72 first, and then take half of that.)

7.   (a) 51% is 50% plus 1%. Fifty percent of 140 is half of 140, which is 70. One percent of 140 is 1.4. So 51% of 140 is 70 plus 1.4, which is 71.4. In equations:

$$
\begin{aligned}
.51 \times 140 &= (.50 + .01) \times 140 \\
&= .50 \times 140 + .01 \times 140 \\
&= 70 + 1.4 = 71.4
\end{aligned}
$$

The distributive property was used at the second equal sign.

(b) 95% is 100% minus 5%. Now 10% of 60 is 6, so 5% of 60 is half of that, which is 3. Therefore 95% of 60 is $60 - 3$, which is 57. In equations:

$$
\begin{aligned}
.95 \times 60 &= (1 - .05) \times 60 \\
&= 1 \times 60 - .05 \times 60 \\
&= 60 - \frac{1}{2} \times .10 \times 60 \\
&= 60 - \frac{1}{2} \times 6 \\
&= 60 - 3 = 57.
\end{aligned}
$$

The distributive property was used at the second equal sign.

(c) Use the fact that $\frac{5}{8} = \frac{4}{8} + \frac{1}{8} = \frac{1}{2} + \frac{1}{8}$.

$$\begin{aligned}
\frac{5}{8} \times 280 &= (\frac{1}{2} + \frac{1}{8}) \times 280 \\
&= \frac{1}{2} \times 280 + \frac{1}{8} \times 280 \\
&= 140 + 35 = 175
\end{aligned}$$

The distributive property was used at the second equal sign.

(d)

$$\begin{aligned}
.99 \times 80 &= (1 - .01) \times 80 \\
&= 1 \times 80 - .01 \times 80 \\
&= 80 - .8 \\
&= 79.2
\end{aligned}$$

The distributive property was used at the second equal sign.

8. (a)

$$\begin{aligned}
25 \times 84 &= 25 \times (4 \times 21) \\
&= (25 \times 4) \times 21 \\
&= 100 \times 21 \\
&= 2100.
\end{aligned}$$

The associative property of multiplication was used to rewrite $25 \times (4 \times 21)$ as $(25 \times 4) \times 21$, in other words, to multiply the 4 with the 25 instead of with the 21.

(b)

$$\begin{aligned}
49 \times 6 &= (50 - 1) \times 6 \\
&= 50 \times 6 - 1 \times 6 \\
&= 300 - 6 \\
&= 294.
\end{aligned}$$

The distributive property was used to rewrite $(50 - 1) \times 6$ as $50 \times 6 - 1 \times 6$.

(c)

$$486 \times 5 \;=\; (243 \times 2) \times 5$$
$$=\; 243 \times (2 \times 5)$$
$$=\; 243 \times 10$$
$$=\; 2430.$$

The associative property of multiplication was used to rewrite $(243 \times 2) \times 5$ as $243 \times (2 \times 5)$.

9. The following sequence of equations shows in full and gory detail why Joey's method is valid and how it uses properties of arithmetic.

$$24 \times 60 \;=\; 24 \times (6 \times 10)$$
$$=\; (24 \times 6) \times 10$$
$$=\; [(25 - 1) \times 6] \times 10$$
$$=\; (25 \times 6 - 1 \times 6) \times 10$$
$$=\; [6 \times 25 - 6] \times 10$$
$$=\; [4 \times 25 + 2 \times 25 - 6] \times 10$$
$$=\; [100 + 50 - 6] \times 10$$
$$=\; (150 - 6) \times 10$$
$$=\; 144 \times 10$$
$$=\; 1440.$$

The associative property of multiplication is used at the 2nd equal sign

The distributive property is used at the 4th and 6th equal signs.

The commutative property of multiplication is used at the 5th equal sign.

The associative property of addition was essentially used at the 8th equal sign.

Problems for Section 4.7 on the Distributive Property

1. Suppose that a child has learned the following basic multiplication facts well:

- The $\times$ *1*, $\times$ *2*, $\times$ *3*, $\times$ *4*, and $\times$ *5* tables, i.e.,

$1 \times 1 = 1$	$1 \times 2 = 2$	$1 \times 3 = 3$	$1 \times 4 = 4$	$1 \times 5 = 5$
$2 \times 1 = 2$	$2 \times 2 = 4$	$2 \times 3 = 6$	$2 \times 4 = 8$	$2 \times 5 = 10$
$3 \times 1 = 3$	$3 \times 2 = 6$	$3 \times 3 = 9$	$3 \times 4 = 12$	$3 \times 5 = 15$
$4 \times 1 = 4$	$4 \times 2 = 8$	$4 \times 3 = 12$	$4 \times 4 = 16$	$4 \times 5 = 20$
$5 \times 1 = 5$	$5 \times 2 = 10$	$5 \times 3 = 15$	$5 \times 4 = 20$	$5 \times 5 = 25$
$6 \times 1 = 6$	$6 \times 2 = 12$	$6 \times 3 = 18$	$6 \times 4 = 24$	$6 \times 5 = 30$
$7 \times 1 = 7$	$7 \times 2 = 14$	$7 \times 3 = 21$	$7 \times 4 = 28$	$7 \times 5 = 35$
$8 \times 1 = 8$	$8 \times 2 = 16$	$8 \times 3 = 24$	$8 \times 4 = 32$	$8 \times 5 = 40$
$9 \times 1 = 9$	$9 \times 2 = 18$	$9 \times 3 = 27$	$9 \times 4 = 36$	$9 \times 5 = 45$

- The squares $1 \times 1 = 1$, $2 \times 2 = 4$, $3 \times 3 = 9, \ldots, 9 \times 9 = 81$.

For each of the following multiplication problems, find *at least 2 different ways* that some of the facts above, together with properties of arithmetic, could be used to mentally calculate the answer to the problem. Explain your answers briefly.

(a) 6×7

(b) 7×8

(c) 6×8

2. Clint and Sue went out to dinner and had a nice meal that cost $64.82. With a 7% tax of $4.54, the total came to $69.36. They want to leave a tip of approximately 15% of the cost of the meal (before the tax). Describe a way that Clint and Sue can *mentally* figure the tip.

3. On a television broadcast on tennis, a commentator said:

   She has made 30 out of 33 first serves. That's better than 90%.

Explain how the commentator could figure this out without using a calculator or doing long division.

4. Your favorite store is having a 10% *off sale*, meaning that the store will take 10% off the price of each item you buy. When the clerk rings up your purchases, she takes 10% off the total, rather than 10% off each item. Will you get the same discount either way? Is there a property of arithmetic related to this? Explain!

5. Show how to use properties of arithmetic in order to calculate 35% of 440 mentally. Explain your strategy and write a string of equations that corresponds to your strategy. Indicate which properties of arithmetic you used, and where. Be specific. Write your equations in the following form:

$$35\% \times 440 \ = \ \text{some expression}$$
$$= \ \vdots$$
$$= \ \text{some expression}$$

6. The following exchanges are taken from *Developing Children's Understanding of the Rational Numbers: A New Model and an Experimental Curriculum* by Joan Moss and Robbie Case [32, page 135]. *"Experimental S1"* and *"Experimental S3"* are two of the fourth grade students who participated in an experimental curriculum described in the article.

Experimenter:	What is 65% of 160?
Experimental S1:	Fifty percent [of 160] is 80. I figure 10%, which would be 16. Then I divided by 2, which is 8 [5%] then 16 plus 8 um ... 24. Then I do 80 plus 24, which would be 104.
$\vdots$	$\vdots$
Experimental S3:	Ten percent of 160 is 16; 16 times 6 equals 96. Then I did 5%, and that was 8, so ... , 96 plus 8 equals 104.

For each of the two students' responses, write strings of equations that correspond to the student's method for calculating 65% of 160. State which properties of arithmetic were used and where (be specific).

Write your string of equations in the following form:

$$65\% \times 160 \ = \ \text{some expression}$$
$$= \ \vdots$$
$$= \ \text{some expression}$$

7. Jenny uses the following method to find 28% of 60,000 mentally:

25% is $\frac{1}{4}$, and $\frac{1}{4}$ of 60 is 15, so 25% of 60,000 is 15,000. One percent of 60,000 is 600, and that times 3 is 1800. So the answer is $15,000 + 1,800$, which is 16,800.

Write a string of equations that calculate 28% of 60,000 and that incorporate Jenny's ideas. Write your equations in the following form:

$$
\begin{aligned}
28\% \times 60,000 \; &= \; \text{some expression} \\
&= \; \text{some expression} \\
&\;\;\vdots \\
&= \; 16,800
\end{aligned}
$$

8. Use the *distributive property* to make it easy for you to calculate 30% of 240 mentally. Then use the *associative property of multiplication* to do the same problem. In each case, explain your strategy in words and then write equations that incorporate your strategy. Write your equations in the following form:

$$
\begin{aligned}
30\% \times 240 \; &= \; \text{some expression} \\
&= \; \text{some expression} \\
&\;\;\vdots
\end{aligned}
$$

9. A sales tax problem:

> Compare the total amount of sales tax you would pay if you went to a store and bought a stereo and an entertainment center at the same time, versus if you first bought the stereo and then went back to the store to buy the entertainment center.

Solve the sales tax problem, then discuss how a property of arithmetic is relevant to comparing the two ways of buying the stereo and entertainment center.

10. (a) Lindsay calculates two fifths of 1260 using the following strategy. First, Lindsay finds half of 1260, which is 630. Then Lindsay subtracts one tenth of 1260, which is 126, from 630 and gives the answer is 504. Discuss the ideas behind Lindsay's strategy. Then

write a string of equations that incorporate Lindsay's strategy, and that show why the strategy is valid. What property of arithmetic is involved? Write your equations in the following form:

$$\frac{2}{5} \times 1260 \;=\; \text{some expression}$$
$$=\; \text{some expression}$$
$$\vdots$$
$$=\; 504$$

(b) Terrell does the same problem in this way: first he multiplies 1260 by 2 to get 2520. Then he multiplies 2520 by 2 to get 5040 and divides this by 10 to get 504. Again, discuss the idea behind Terrell's strategy. Then write a string of equations that incorporate Terrell's strategy, and that show why the strategy is valid. Write your equations in the form shown above.

11. Frank's Jewelers runs the following advertisement: "Come to our 40% off sale on Saturday. We're not like the competition, who raise prices by 30% and then have a 70% off sale!"

(a) If two items start off with the same price, which gives you the lower price in the end: taking off 40% or raising the price by 30% and then taking off 70% (of the raised price)?

(b) Consider the same problem more generally, with other numbers. For example, if you raise prices by 20% and then take off 50% (of the raised price) how does this compare to taking 30% off of the original price? If you raise a price by 30% and then lower the raised price by 30%, how does that compare to the original price? Try at least 2 other pairs of percentages by which to raise and then lower a price. Describe what you observe.

Predict what happens in general: if you raise a price by $A\%$ and then take $B\%$ off of the raised price, does that have the same result as if you'd lowered the original price by $(B - A)\%$? If not, which produces the lower final price?

(c) Use the distributive property or FOIL to explain the pattern you discovered in part (b). Remember that to raise a price by 15%, for example, you multiply the price by $1 + .15$, while to lower a price by 15%, you multiply the price by $1 - .15$.

12. (a) Use an ordinary calculator to calculate $666, 666, 666 \times 999, 999, 999$. Based on the calculator's display, guess how to write the answer in ordinary decimal notation (showing *all* digits).

    (b) Now think some more, and determine how to write the product $666, 666, 666 \times 999, 999, 999$ in ordinary decimal notation, showing all its digits, *without* multiplying longhand or using a calculator or computer.

13. Calculate the product $9, 999, 999, 999 \times 9, 999, 999, 999$ *without* using a calculator or computer. Give the answer in ordinary decimal notation, showing all its digits. (Both numbers in the product have 10 nines.)

14. Check that:

    $11 - 2 = 3^2$

    $1111 - 22 = 33^2$

    $111111 - 222 = 333^2$.

    Continue to find at least three more in the pattern. Does the pattern continue? Now explain why there is such a pattern. Hint: notice that $1111 - 22 = 11 \times 101 - 11 \times 2$.

15. Determine which of the following two numbers is larger and explain your reasoning:

    (a) $1, 000, 000 \times (1 + 2 + 3 + 4 + \cdots + 1, 000, 001)$ or

    (b) $1, 000, 001 \times (1 + 2 + 3 + 4 + \cdots + 1, 000, 000)$

16. The **square** of a number is just the number times itself. For example, the square of 4 is 16.

    (a) Find the squares of the numbers $15, 25, 35, 45, \ldots, 95, 105, 115, 125, \ldots, 195, 205$, and three other whole numbers that end in 5.

    (b) Find some patterns in the answers to part (a). Specifically, in all of your answers to part (a), what do you notice about the last two digits, and what do you notice about the number formed by deleting the last two digits?

(c) Use what you discovered in part (b) to predict the squares of 2,005 and 10,005.

(d) Every whole number ending in 5 must be of the form $10A + 5$ for some whole number A. Find the square of $10A + 5$, namely calculate $(10A + 5) \times (10A + 5)$.

(e) How does your answer to part (d) explain the pattern you found in part (b)?

17. Try out this mathematical magic trick. On a piece of paper:

i) Write the number of days a week you would like to go out (From 1 to 7)

ii) Multiply the number by 2

iii) Add 5

iv) Multiply by 50

v) If you have already had your birthday this year, add 1752 if it is the year 2002 (add 1753 if it is 2003, add 1754 if it is 2004, add 1755 if it is 2005, and so on). If you have not yet had your birthday this year, add 1751 if it is the year 2002 (add 1752 if it is 2003, add 1753 if it is 2004, add 1754 if it is 2005, and so on).

vi) Last step: subtract the four-digit year you were born. You should now have a three digit number. If not, try again and don't peek below! If you have a three digit number, continue:

The first digit of your answer is your original number (i.e., how many times you want to go out each week). The second two digits are your current age.

Is it magic, or is it math? Explain why the trick works!

18. In order to subtract 197 from 384 you need to regroup. The following string of equations correspond to the regrouping process, when it is shown in gory detail with expanded forms:

$$
\begin{aligned}
384 \;&=\; 3(100) + 8(10) + 4(1) && (4.1)\\
&=\; [2(100) + 1(100)] + [7(10) + 1(10)] + 4(1) && (4.2)\\
&=\; [2(100) + 10(10)] + [7(10) + 10(1)] + 4(1) && (4.3)
\end{aligned}
$$

$$= \quad [2(100) + 10(10)] + 7(10) + [10(1) + 4(1)] \qquad (4.4)$$
$$= \quad 2(100) + [10(10) + 7(10)] + [10(1) + 4(1)] \qquad (4.5)$$
$$= \quad 2(100) + 17(10) + 14(1) \qquad (4.6)$$

Which properties of arithmetic are used, and where? Be thorough and be specific!

19. In order to subtract .36 from .84 you need to regroup. The following string of equations correspond to the regrouping process, when it is shown in gory detail with expanded forms:

$$.84 \quad = \quad 8\left(\frac{1}{10}\right) + 4\left(\frac{1}{100}\right) \qquad (4.7)$$

$$= \quad [7\left(\frac{1}{10}\right) + 1\left(\frac{1}{10}\right)] + 4\left(\frac{1}{100}\right) \qquad (4.8)$$

$$= \quad [7\left(\frac{1}{10}\right) + 10\left(\frac{1}{100}\right)] + 4\left(\frac{1}{100}\right) \qquad (4.9)$$

$$= \quad 7\left(\frac{1}{10}\right) + [10\left(\frac{1}{100}\right) + 4\left(\frac{1}{100}\right)] \qquad (4.10)$$

$$= \quad 7\left(\frac{1}{10}\right) + 14\left(\frac{1}{100}\right) \qquad (4.11)$$

Which properties of arithmetic are used in this, and where? Be thorough and be specific!

4.8   Why the Procedure for Multiplying Whole Numbers Works

The standard longhand procedure for multiplying multiple-digit whole numbers is an efficient paper and pencil method of calculation. This method is useful because it converts a multiplication problem with numbers that have *several digits* to many multiplication problems with *one-digit* numbers. Therefore, someone who has memorized the one-digit multiplication tables (from $1 \times 1 = 1$ to $9 \times 9 = 81$) can multiply any pair of whole numbers by using the longhand multiplication algorithm. But why does this clever method give the correct answer to multiplication problems? What makes it work? This is what we will examine in this section.

Before we continue, let's first make sense of the questions at the end of the previous paragraph. Notice the distinction between the *meaning* of multiplication and the longhand *procedure* for multiplying. If you have 58 bags of widgets and there are 764 widgets in each bag, then you can ask how many widgets you have in all. There is some specific total number of widgets. But what is this number? According to the *meaning of multiplication*, the total number of widgets is

$$58 \times 764.$$

But the longhand multiplication *procedure* consists of carrying out a number of steps:

> multiply 8×4, write the 2, carry the 3; multiply 8×6, add the carried 3, write the 1, carry the 5, etc.

What is the connection between the *meaning of* 58×764 and the *procedure for calculating* 58×764? Or:

> *why does the standard longhand multiplication procedure for calculating* 58×764 *give the actual total number of widgets in 58 bags if there are 764 widgets in each bag?*

We will explain this in this section.

The Partial Products Multiplication Algorithm

In order to understand the standard longhand multiplication algorithm, we will work with the **partial products multiplication algorithm**. The partial products algorithm is essentially the same as the standard multiplication algorithm, but it has the virtue of showing more steps, so it will make our analysis easier. A few examples will illustrate how the partial products algorithm works. In the examples below, the arrows and the expressions to the right of the arrows have been added to show how to carry out the algorithm. You do not have to write these arrows and expressions when you use the algorithm yourself.

Standard Algorithm :   Partial Products Algorithm :

```
    4
   38                      38
  ×6                      ×6
  228                      48   ← 6 × 8
                          180   ← 6 × 30
                          228   ← add
```

Standard Algorithm :   Partial Products Algorithm :

```
   2 1
  274                     274
  ×3                      ×3
  822                      12   ← 3 × 4
                          210   ← 3 × 70
                          600   ← 3 × 200
                          822   ← add
```

StandardAlgorithm :   PartialProductsAlgorithm :

```
    1
    1
   45                       45
  ×23                      ×23
  135                       15   ← 3 × 5
  900                      120   ← 3 × 40
 1035                      100   ← 20 × 5
                           800   ← 20 × 40
                          1035   ← add
```

Class Activity 4T: The Standard Versus the Partial Products Multiplication Algorithm

Class Activity 4U: A First Look at Why the Partial Products Algorithm Gives Correct Answers

Class Activity 4V: Another Look at Why the Partial Products Multiplication Algorithm Gives Correct Answers

The Standard and Partial Products Multiplication Algorithms and Why They Give Correct Answers

As at the beginning of the section, consider the situation of 58 bags of widgets with 764 widgets in each bag. Our goal is to explain why the partial products multiplication algorithm, and therefore also the standard multiplication algorithm, correctly calculates the total number of widgets.

Comparing the Two Algorithms

Compare the way the standard and partial products algorithms are used to calculate 58×764:

Standard Algorithm :

$$
\begin{array}{r}
{\scriptstyle 3\,2} \\
{\scriptstyle 5\,3} \\
764 \\
\times 58 \\
\hline
\end{array}
$$

$6112 \leftarrow 32 + 480 + 5600$

$\underline{38200} \leftarrow 200 + 3000 + 35000$

44312

Partial Products Algorithm :

$$
\begin{array}{r}
764 \\
\times 58 \\
\hline
\end{array}
$$

$$
\left\{
\begin{array}{ll}
32 & \leftarrow 8 \times 4 \\
480 & \leftarrow 8 \times 60 \\
5600 & \leftarrow 8 \times 700
\end{array}
\right.
$$

$$
\left\{
\begin{array}{ll}
200 & \leftarrow 50 \times 4 \\
3000 & \leftarrow 50 \times 60 \\
\underline{35000} & \leftarrow 50 \times 700
\end{array}
\right.
$$

44312

Notice that the standard algorithm *condenses the steps in the partial products algorithm.* In this example, three lines of calculations in the partial products algorithm are condensed to one line in the standard algorithm: the 6112 produced in the standard algorithm can be obtained by adding the three lines 8×4, 8×60, and 8×700 of the partial products algorithm; the 38200 can be obtained by adding 50×4, 50×60, and 50×700. When you multiply and then add carried numbers in the standard algorithm, you are really just *combining lines from the partial products algorithm.*

Although the partial products algorithm is longer than the standard algorithm, on the whole it is easier to carry out. However, one way that the partial products algorithm is more difficult is in having to pay attention to place value. In the standard algorithm you don't have to pay too much attention to place value—as long as you line up the numbers carefully, the algorithm takes care of place value, and you only have to multiply pairs of *one-digit* numbers. In the partial products algorithm, on the other hand, you do have to pay attention to place value. For example, in the last line, you have to be able to calculate 50×700, instead of just 5×7. Notice that relating 50×700 to 5×7 uses the commutative property of multiplication:

$$
\begin{aligned}
50 \times 700 \ &= \ 5 \times 10 \times 7 \times 100 \\
&= \ 5 \times 7 \times 10 \times 100 \text{ (by the commutative property of multiplication)} \\
&= \ 35 \times 1000 \\
&= \ 35000
\end{aligned}
$$

The associative property of multiplication is used tacitly at the third equal sign because the 10 is multiplied with the 100 instead of with the 5×7.

Why We Place Extra Zeros on Some Lines in the Standard Algorithm

We can use the partial products algorithm to explain why it makes sense that we put an initial zero in the ones column of the second line in the standard multiplication algorithm (in the example above, this is the line produced by 5×4, 5×6 and 5×7). As the partial products algorithm reveals, that line is actually produced by *multiplying by 50 rather than by 5*. By placing the initial zero in the ones place, all digits are moved one place to the left, which has the same effect as multiplying by 10. The net effect is that this second line is now actually 50 times 764 rather than 5 times 764.

What if the multiplication problem was 258×764 instead of 58×764? Then in the standard algorithm there would be a third line produced by 2×4, 2×6, and 2×7. Before beginning the calculations for this third line, you would put down *two zeros*. Why two zeros? Because this line should really be produced by multiplying by 200 instead of by 2. By placing zeros in the ones and tens places, all digits are moved two places to the left, which has the same effect as multiplying by 100. The net effect is that the third line is then 200 times 764 instead of 2 times 764.

Why the Algorithms Give Correct Answers

The key to explaining why the partial product algorithm, and therefore also the standard algorithm, gives the correct answer to 58×764, lies in observing that *each step in the alogrithm counts the number of widgets in a portion* of the 58 bags of 764 widgets. The algorithms come from a clever way of subdividing the full collection of widgets into smaller portions whose numbers are easy to calculate as long as you know your one-digit multiplication tables. At the end of the algorithm, when you add the lines produced by multiplying, you are really just adding the numbers of widgets in each portion to get the total number of widgets.

Below is a diagram indicating the total number of widgets in 58 rows with 764 widgets in each row. Each *o* in the diagram represents a widget. Because the numbers 58 and 764 are so large, the diagram doesn't actually show all 58 rows, and it doesn't show all 764 widgets in each row.

	700 widgets	60 widgets	4 widgets
	OOOOOOOOOO ... OOOOOOOOOO	OOOOOOOOOO ... OOOOOOOOOO	OOOO
	OOOOOOOOOO ... OOOOOOOOOO	OOOOOOOOOO ··· OOOOOOOOOO	OOOO
	OOOOOOOOOO ... OOOOOOOOOO	OOOOOOOOOO ··· OOOOOOOOOO	OOOO
	OOOOOOOOOO ... OOOOOOOOOO	OOOOOOOOOO ··· OOOOOOOOOO	OOOO
	OOOOOOOOOO ... OOOOOOOOOO	OOOOOOOOOO ··· OOOOOOOOOO	OOOO
50 rows	50×700 widgets here	50×60 widgets here	50×4 widgets here
	OOOOOOOOOO ... OOOOOOOOOO	OOOOOOOOOO ··· OOOOOOOOOO	OOOO
	OOOOOOOOOO ... OOOOOOOOOO	OOOOOOOOOO ··· OOOOOOOOOO	OOOO
	OOOOOOOOOO ... OOOOOOOOOO	OOOOOOOOOO ··· OOOOOOOOOO	OOOO
	OOOOOOOOOO ... OOOOOOOOOO	OOOOOOOOOO ··· OOOOOOOOOO	OOOO
	OOOOOOOOOO ... OOOOOOOOOO	OOOOOOOOOO ··· OOOOOOOOOO	OOOO
	OOOOOOOOOO ... OOOOOOOOOO	OOOOOOOOOO ... OOOOOOOOOO	OOOO
	OOOOOOOOOO ... OOOOOOOOOO	OOOOOOOOOO ··· OOOOOOOOOO	OOOO
	OOOOOOOOOO ... OOOOOOOOOO	OOOOOOOOOO ··· OOOOOOOOOO	OOOO
8 rows	8×700 widgets here	8×60 widgets here	8×4 widgets here
	OOOOOOOOOO ... OOOOOOOOOO	OOOOOOOOOO ··· OOOOOOOOOO	OOOO
	OOOOOOOOOO ... OOOOOOOOOO	OOOOOOOOOO ··· OOOOOOOOOO	OOOO
	OOOOOOOOOO ... OOOOOOOOOO	OOOOOOOOOO ··· OOOOOOOOOO	OOOO

The diagram above illustrates how to subdivide the full collection of widgets so that the number of widgets in each portion corresponds to a line in the partial products algorithm. So, for instance, the line 5600 in the partial products algorithm, which comes from 8 × 700, is the number of widgets in the piece in the lower left-hand corner which represents 8 groups of 700 widgets. Similarly, the line 200, which comes from 50 × 4, is the number of widgets in the piece at the top right-hand corner which represents 50 groups of 4 widgets. It's the same for all the other lines in the partial products algorithm—*each line in the algorithm corresponds to a portion of the collection of widgets*. So when you add up the lines, you get the total number of widgets. It's a very clever method!

Another way to explain why the partial products algorithm, and therefore also the standard algorithm, gives the correct answer to 58×764 is to put the numbers 58 and 764 in expanded forms and use properties of arithmetic to calculate 58×764:

$$
\begin{aligned}
58 \times 764 &= (50 + 8) \times (700 + 60 + 4) \\
&= 50 \times (700 + 60 + 4) + 8 \times (700 + 60 + 4) \\
&= 50 \times 700 + 50 \times 60 + 50 \times 4 + 8 \times 700 + 8 \times 60 + 8 \times 4 \\
&= 35{,}000 + 3{,}000 + 200 + 5{,}600 + 480 + 32 \\
&= 44{,}312
\end{aligned}
$$

As the equations above show, the six lines produced by the partial products algorithm are exactly the six products

$$50 \times 700, \ 50 \times 60, \ 50 \times 4, \ 8 \times 700, \ 8 \times 60, \ 8 \times 4,$$

produced by using the distributive property several times, starting with

$$(50 + 8) \times (700 + 60 + 4).$$

Class Activity 4W: Correcting a Multiplication Misconception

Class Activity 4X: 29×31 Versus 30×30

Class Activity 4Y: Squares of Some Numbers Ending in 1

Exercises for Section 4.8 on the Multiplication Algorithms

1. Find the last two digits (the ones and hundreds digits) of

$$798{,}312{,}546{,}936 \times 74.$$

2. Explain why the standard multiplication algorithm gives the correct answer to the multiplication problem $\begin{array}{r} 45 \\ \times 23 \\ \hline \end{array}$ Be sure to explain why we put a 0 in the second row of multiplication (the one produced by multiplying by 2).

3. Cameron wants to calculate 23×23. She says: $20 \times 20 = 400$ and $3 \times 3 = 9$, so $23 \times 23 = 400 + 9 = 409$. Cameron has a good idea, but her reasoning is not completely correct. How could Cameron adjust her reasoning to get the correct answer? Help her with a picture!

Then use *properties of arithmetic* or FOIL to calculate 23×23. Relate this calculation to the partial products and standard multiplication algorithms.

Answers to Exercises for Section 4.8 on the Multiplication Algorithms

1. The last two digits are 64. Notice that you only have to find the last two digits of 36×74 to find this. Why? Think about doing longhand multiplication: any other contribution will be in the thousands place or higher.

2. The standard multiplication algorithm is basically just a condensed version of the partial products algorithm, so it will be enough to explain why the partial products algorithm works. Class Activity 4V outlines several ways to explain why the partial products algorithm works.

Using a diagram: the four lines produced by the partial products algorithm correspond to four parts in the diagram shown in the class activity, namely 20 rows of 40, 20 rows of 5, 3 rows of 40, and 3 rows of 5. See Figure 4.29. Because the picture shows 23 rows with 45 "*o*"s in each row, it has

$$23 \times 45$$

"*o*"s, according to the meaning of multiplication. In the four parts there are a total of

$$20 \times 40 + 20 \times 5 + 3 \times 40 + 3 \times 5$$

"*o*"s. Since it's the same number of "*o*"s no matter how you count them, therefore

$$23 \times 45 = 20 \times 40 + 20 \times 5 + 3 \times 40 + 3 \times 5.$$

The four terms 20×40, 20×5, 3×40, 3×5, are exactly the four lines produced by the partial products algorithm, therefore Figure 4.29 helps to show why the algorithm produces the correct answer.

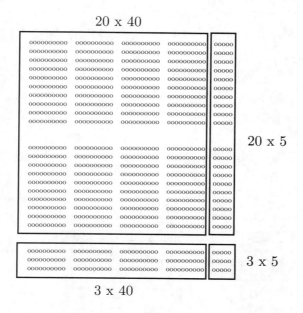

Figure 4.29: Showing $23 \times 45 = 20 \times 40 + 20 \times 5 + 3 \times 40 + 3 \times 5$

Using expanded forms and properties of arithmetic:

$$
\begin{aligned}
23 \times 45 &= (20 + 3) \times (40 + 5) \\
&= 20 \times (40 + 5) + 3 \times (40 + 5) \\
&= 20 \times 40 + 20 \times 5 + 3 \times 40 + 3 \times 5
\end{aligned}
$$

The distributive property was used at the second and third equal signs. The four terms 20×40, 20×5, 3×40, 3×5, are exactly the four lines produced by the partial products algorithm, therefore the partial products algorithm is really just a way of displaying the results produced by using the distributive property when the numbers 23 and 45 are put in expanded form.

See the text for why a zero must be placed in the second line of the standard algorithm.

3. Figure 4.30 shows 23 rows of 23 small squares, subdivided into four portions. The picture shows that Cameron has left out some portions in her calculation. She needs to add 20×3 and 3×20 (the darkly shaded portions). In terms of properties of arithmetic:

$$
\begin{aligned}
23 \times 23 &= (20 + 3) \times (20 + 3) \\
&= 20 \times (20 + 3) + 3 \times (20 + 3) \\
&= 20 \times 20 + 20 \times 3 + 3 \times 20 + 3 \times 3
\end{aligned}
$$

The distributive property was used at the second and third equal signs.

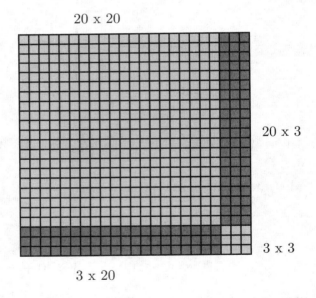

20 x 20

20 x 3

3 x 3

3 x 20

Figure 4.30: $23 \times 23 = 20 \times 20 + 20 \times 3 + 3 \times 20 + 3 \times 3$

Problems for Section 4.8 on the Multiplication Algorithms

1. Solve the multiplication problem

$$237 \\ \underline{\times 43}$$

in three different ways: using the standard algorithm, using the partial products algorithm, and by writing the numbers in expanded forms and using properties of arithmetic. For each of the three methods, discuss how the steps in that method are related to the steps in the other methods.

2. Ted tries to calculate 13×14 by using the following method:

$$
\begin{aligned}
13 \times 14 &= 10 \times 10 + 3 \times 4 \\
&= 100 + 12 \\
&= 112.
\end{aligned}
$$

   (a) Help Ted see that he has a good idea but that his method is not yet correct. Explain how Ted could modify his method to get the correct answer. Draw a picture to aid your explanations.

   (b) Write equations that incorporate Ted's idea but that compute 13×14 correctly.

3. SunJae is working on the multiplication problem 21×34. SunJae says that he can take 1 from the 21 and put it with the 34 to get 20×35. SunJae says that this new multiplication problem, 20×35, should have the same answer as 21×34. Is SunJae right? How might you convince SunJae that his reasoning is or is not correct in a way *other than simply showing him the answers to the two multiplication problems*?

4. Maria is working on the multiplication problem 38×25. Maria says:

   4 times 25 is 100, so 40 times 25 is 1000. Now take away 2, so the answer is 998.

   Is Maria's method correct or not? If it is correct, write equations that incorporate Maria's work and that show why it's correct. If Maria's reasoning is not correct, then *working with portions that are right*, correct

Maria's work and write a string of equations equations that correspond to your corrected method for calculating 38×25. Write your equations in the following form:

$$38 \times 25 \quad = \quad \text{some expression}$$
$$= \quad \text{some expression}$$
$$\vdots$$

5. TreVon wants to calculate 4030×5020. He says:

$40 \times 50 = 2000$ and $30 \times 20 = 600$, so $4030 \times 5020 = 2000 + 600 = 2600$.

But then TreVon realizes that his answer can't be right because it's too small. How could TreVon *adjust* his reasoning to get the correct answer? Draw a supporting picture. (Don't just start over in a different way, work with TreVon's ideas and modify them to get correct reasoning.) Then write corresponding equations that use *properties of arithmetic* to calculate 4030×5020. Write your equations in the following form:

$$4030 \times 5020 \quad = \quad \text{some expression}$$
$$= \quad \text{some expression}$$
$$\vdots$$

6. There is an interesting mental technique for multiplying certain pairs of numbers. A few of examples will illustrate how it works:

- To calculate 32×28, notice that the two factors 28 and 32 are both 2 away—in opposite directions—from 30. To calculate 32×28, do the following:

$$30 \times 30 - 2 \times 2 \quad = \quad 900 - 4$$
$$= \quad 896.$$

Notice that you can do this calculation in your head.

- Similarly, to calculate 59×61, notice that both factors are 1 away—in opposite directions—from 60. Then 59×61 is

$$60 \times 60 - 1 \times 1 \ = \ 3600 - 1$$
$$= \ 3599.$$

Once again, notice that you can do this mentally.

(a) Use the method above to calculate 83×77, and 195×205, and one other multiplication problem like this that you make up.

(b) Now explain why this method works. (*Hint:* A diagram might be helpful. Another approach is to notice that the technique applies to multiplication problems of the form $(A + B) \times (A - B)$.)

7. The **square** of a number is just the number times itself. For example, the square of 4 is 16.

(a) Calculate the squares of $1, 2, \ldots, 9$ and many other whole numbers including $17, 34, 61, 82, 99, 123, 255, 386, 728$. Record the ones digits in each case. What do you notice? Do any of your squares have a ones digit of 7, for example? Are any other digits missing from the ones digits of squares?

(b) Which digits can never occur as the ones digit of a square of a whole number? Explain why some digits cannot occur as the ones digit of a square of a whole number.

(c) Based on what you've discovered, could the number

$$139, 787, 847, 234, 329, 483$$

be the square of a whole number? Why or why not?

8. Consider the product $127, 500, 000, 002 \times 355, 700, 000, 003$.

(a) Sally said that the digit in the one hundred millions place of this product is the ones digit of the number $5 \times 3 + 2 \times 7$, which is 9. Explain how you can tell that Sally is right without actually multiplying the numbers out completely.

(b) What if the multiplication problem had been $127, 598, 765, 432 \times 355, 789, 012, 343$. Would Sally's reasoning still apply to find the digit in the hundred millions place? Why or why not?

4.9   Multiplying Fractions

In this section, we will study the meaning of multiplication of fractions and we will explain why the standard *procedure* for multiplying fractions gives gives answers to fraction multiplication problems that agree with what we expect from the *meaning* of multiplication.

The Meaning of Multiplication for Fractions

At the beginning of this chapter we defined multiplication as follows: If A and B are non-negative numbers, then

$$A \times B \quad \text{or} \quad A \cdot B$$

means the total number of objects in A groups if there are B objects in each group. This definition applies not only to whole numbers, but also to fractions and decimals. But as we'll see, when fractions are involved, we may prefer to reword the definition for the sake of clarity.

Consider the following examples:

1. $\frac{1}{2} \times 21$ means the total number of objects in $\frac{1}{2}$ groups if there are 21 objects in each group.

   Better wording: $\frac{1}{2} \times 21$ means the total number of objects in $\frac{1}{2}$ *of a group* if there are 21 objects in each group.

2. $\frac{1}{2} \times \frac{1}{2}$ means the total number of objects in $\frac{1}{2}$ groups if there are $\frac{1}{2}$ objects in each group.

   Better wording: $\frac{1}{2} \times \frac{1}{2}$ means the total number of objects in $\frac{1}{2}$ *of a group* if there *is $\frac{1}{2}$ of an object* in each group.

3. $3 \times \frac{2}{5}$ means the total number of objects in 3 groups if there are $\frac{2}{5}$ objects in each group. (See Figure 4.31.)

   Better wording: $3 \times \frac{2}{5}$ means the total number of objects in 3 groups if there *is $\frac{2}{5}$ of an object* in each group.

What do we mean by the word *object*? In a mathematical context, the word *object* can refer not just to a single thing, but also to a collection of things (as in "the cars in the US"), or to a quantity (as in "one cup of water"

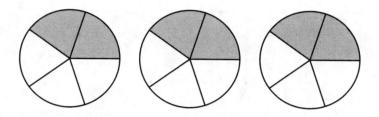

Figure 4.31: $3 \times \frac{2}{5}$

or "\$20,000"). In the context of fractions, it is also natural to use the term *a whole* instead of *object*.

The wording "*the total number of objects*" must be understood to include the possibility of fractional objects. For example, if the object we're considering is a full jar of peanut butter, then what does the following mean?

> The total number of objects in $\frac{1}{2}$ of a group when each group contains $\frac{1}{2}$ of a full jar of peanut butter.

In this case, what we mean by *the total number of objects* is really *the fraction of a full jar of peanut butter* in $\frac{1}{2}$ *of* $\frac{1}{2}$ of a full jar of peanut butter. This suggests alternate wording for the meaning of fraction multiplication, which you can use if you find it clearer. It is most appropriate in cases where both factors are proper fractions:

> $\frac{A}{B} \times \frac{C}{D}$ means the fraction of an object in $\frac{A}{B}$ of $\frac{C}{D}$ of the object.

For example:

> $\frac{1}{2} \times \frac{3}{4}$ means the fraction of a container of yogurt in $\frac{1}{2}$ of $\frac{3}{4}$ of the container of yogurt. (See Figure 4.32.)

Class Activity 4Z: When Do We Multiply Fractions?

The Procedure for Multiplying Fractions

The procedure for multiplying fractions is very easy: just multiply the numerators and multiply the denominators, in other words:

$$\frac{A}{B} \cdot \frac{C}{D} = \frac{A \cdot C}{B \cdot D}.$$

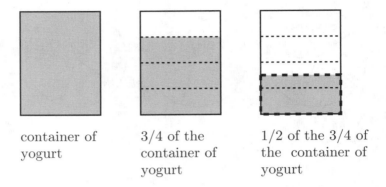

container of   3/4 of the   1/2 of the 3/4 of
yogurt   container of   the container of
                      yogurt   yogurt

Figure 4.32: $\frac{1}{2}$ of $\frac{3}{4}$ of a Container of Yogurt

For example,

$$\frac{2}{3} \cdot \frac{4}{5} = \frac{2 \cdot 4}{3 \cdot 5} = \frac{8}{15}.$$

Notice that the procedure can be used even when whole numbers are involved because a whole number can always be written as a fraction by "putting it over 1". For example,

$$\frac{2}{3} \cdot 17 = \frac{2}{3} \cdot \frac{17}{1} = \frac{2 \cdot 17}{3 \cdot 1} = \frac{34}{3} = 11\frac{1}{3}.$$

As with the procedure for multiplying whole numbers, we want to explain why this procedure gives correct answers to fraction multiplication problems. The meaning of fractions and the meaning of multiplication tell us what an expression of the form

$$\frac{A}{B} \cdot \frac{C}{D}$$

means. We must explain *why this is equal to the fraction*

$$\frac{A \cdot C}{B \cdot D}.$$

Class Activity 4AA: A First Look at Explaining Why the Procedure for Multiplying Fractions Gives Correct Answers

Class Activity 4BB: Misconceptions With Fraction Multiplication

Class Activity 4CC: Explaining Why the Procedure for Multiplying Fractions Gives Correct Answers

Explaining Why the Procedure for Multiplying Fractions Gives Correct Answers

The procedure for multiplying fractions is easy to carry out. It also seems sensible because it involves multiplying the numerators and the denominators. However, remember that we *don't* add fractions by simply *adding* the numerators and *adding* the denominators. So why is the simple procedure of multiplying the numerators and multiplying the denominators valid for fraction multiplication? We will use the meaning of fractions, the meaning of multiplication, and logical reasoning to explain why the simple procedure for multiplying fractions gives correct answers to fraction multiplication problems.

Consider the multiplication problem

$$\frac{2}{7} \cdot \frac{3}{4},$$

and consider $\frac{2}{7} \cdot \frac{3}{4}$ of a rectangle. According to the meaning of multiplication:

> $\frac{2}{7} \cdot \frac{3}{4}$ means the fraction of a rectangle in $\frac{2}{7}$ of a group if there is $\frac{3}{4}$ of a rectangle in each group.

> In other words: $\frac{2}{7} \cdot \frac{3}{4}$ means the fraction of a rectangle in $\frac{2}{7}$ of $\frac{3}{4}$ of the rectangle.

So starting with a rectangle, first consider $\frac{3}{4}$ of a rectangle, and then consider $\frac{2}{7}$ of the $\frac{3}{4}$ of the rectangle, as shown in Figure 4.33.

Here is the crucial point: we must identify the $\frac{2}{7}$ of the $\frac{3}{4}$ of the rectangle *as a fraction of the original rectangle*. To identify what fraction of the original rectangle this $\frac{2}{7}$ of the $\frac{3}{4}$ of the rectangle is, we must put it back inside the original rectangle, as shown in Figure 4.34.

Because the original rectangle was first divided into 4 equal parts, and because each of those parts was then divided into 7 equal parts, therefore the original rectangle has been subdivided into a total of $7 \cdot 4$ small equal

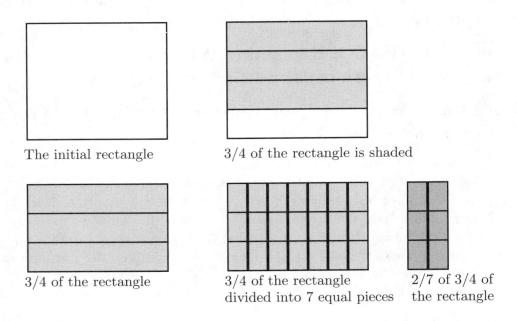

The initial rectangle   3/4 of the rectangle is shaded

3/4 of the rectangle   3/4 of the rectangle   2/7 of 3/4 of
                              divided into 7 equal pieces   the rectangle

Figure 4.33: $\frac{2}{7}$ of $\frac{3}{4}$ of a Rectangle

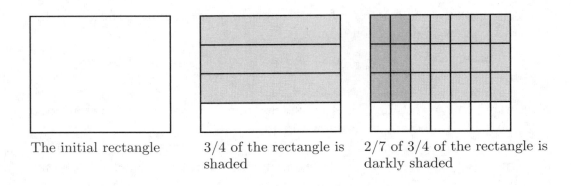

The initial rectangle   3/4 of the rectangle is   2/7 of 3/4 of the rectangle is
                               shaded   darkly shaded

Figure 4.34: What Fraction is $\frac{2}{7}$ of $\frac{3}{4}$ of a Rectangle?

parts. Notice that we *multiply* because there are 7 columns with 4 pieces in each column. Of those $7 \cdot 4$ equal parts, the darkly shaded parts (which represent our $\frac{2}{7}$ of the $\frac{3}{4}$ of the rectangle) form $2 \cdot 3$ parts. Once again, notice that we must *multiply* because there are 2 columns with 3 small pieces in each column that are darkly shaded. So the $\frac{2}{7}$ of the $\frac{3}{4}$ of the rectangle is represented by $2 \cdot 3$ parts out of a total of $7 \cdot 4$ equal parts making up the original rectangle. This means that $\frac{2}{7}$ of the $\frac{3}{4}$ of the rectangle is

$$\frac{2 \cdot 3}{7 \cdot 4}$$

of the rectangle. Therefore

$$\frac{2}{7} \cdot \frac{3}{4} = \frac{2 \cdot 3}{7 \cdot 4},$$

in other words, the *procedure* for multiplying fractions gives the answer we expect from the *meaning* of fractions and the *meaning* of multiplication.

There wasn't anything special about the numbers 2, 3, 7, and 4 that were used here, any other natural numbers could have been substituted and the argument above would work in the same way. This explains why we multiply fractions by multiplying the numerators and multiplying the denominators, in other words it gives us a good reason for why we multiply fractions the way we do:

$$\frac{A}{B} \cdot \frac{C}{D} = \frac{A \cdot C}{B \cdot D}.$$

Why Do We Need to Know This?

As we've seen, the *procedure* for multiplying fractions is easy, but the explanation for why this procedure gives correct answers to fraction multiplication problems is much harder. As a teacher, it is especially important for you to have a deep *conceptual* understanding of the mathematics you will teach. This means not just knowing *how*, but also *why* and *when* the various procedures and formulas in mathematics work. Why is this important? Research shows that teachers who have a conceptual understanding of mathematics tend to use a conceptually directed teaching strategy (see [29, especially pages 38–39]). These teachers teach in a way that encourages understanding and sense-making. Research also shows that this requires deep subject matter knowledge:

"Limited subject matter knowledge restricts a teacher's capacity to promote conceptual learning among students. Even a strong belief of "teaching mathematics for understanding" cannot remedy or supplement a teacher's disadvantage in subject matter knowledge. A few beginning teachers in the procedurally directed group wanted to "teach for understanding." They intended to involve students in the learning process, and to promote conceptual learning that explained the rationale underlying the procedure. However, because of their own deficiency in subject matter knowledge, their conception of teaching could not be realized." [29, page 36])

Exercises for Section 4.9 on Multiplying Fractions

1. What does $\frac{3}{4} \times \frac{1}{2}$ mean? Give an example of a story problem that is solved by calculating $\frac{3}{4} \times \frac{1}{2}$.

2. Which of the following problems are solved by calculating $\frac{1}{3} \times \frac{2}{3}$ and which are not?

   (a) A recipe calls for $\frac{2}{3}$ of a cup of sugar. You want to make $\frac{1}{3}$ of the recipe. How much sugar should you use?

   (b) $\frac{2}{3}$ of the cars at a car dealership have power steering. $\frac{1}{3}$ of the cars at the same car dealership have side mounted airbags. What fraction of the cars at the car dealership have both power steering and side mounted airbags?

   (c) $\frac{2}{3}$ of the cars at a car dealership have power steering. $\frac{1}{3}$ of those cars that have power steering have side mounted airbags. What fraction of the cars at the car dealership have both power steering and side mounted airbags?

   (d) Ed put $\frac{2}{3}$ of a bag of candies in a batch of cookies that he made. Ed ate $\frac{1}{3}$ of the batch of cookies. How many candies did Ed eat?

   (e) Ed put $\frac{2}{3}$ of a bag of candies in a batch of cookies that he made. Ed ate $\frac{1}{3}$ of the batch of cookies. What fraction of a bag of candies did Ed eat?

3. Use the *meaning* of fractions and the *meaning* of multiplication to explain why
$$\frac{2}{7} \cdot \frac{3}{4} = \frac{2 \cdot 3}{7 \cdot 4}.$$

4. Anthony is trying to solve $\frac{5}{7} \cdot 2$. He draws a picture as in Figure 4.35 and concludes from his picture that $\frac{5}{7} \cdot 2 = \frac{10}{14}$ because there are 14 small pieces and 10 of them are shaded. Is Anthony right or not? If not, help Anthony figure out what's wrong with his reasoning.

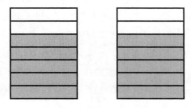

Figure 4.35: Is $\frac{5}{7} \cdot 2 = \frac{10}{14}$?

5. Maryann has to calculate $\frac{2}{3} \times 6\frac{3}{4}$. Rather than convert the $6\frac{3}{4}$ to an improper fraction, she solves the problem this way. She finds $\frac{2}{3}$ of 6, which is 4, and she finds $\frac{2}{3}$ of $\frac{3}{4}$, which is $\frac{1}{2}$. Then Maryann says the answer is $4\frac{1}{2}$. Write equations that correspond to Maryann's work and that explain why her work is correct. Which properties of arithmetic are involved?

Answers to Exercises for Section 4.9 on Multiplying Fractions

1. $\frac{3}{4} \times \frac{1}{2}$ means the fraction of an object in $\frac{3}{4}$ of a group of $\frac{1}{2}$ of the object. A sample story problem: A recipe calls for $\frac{1}{2}$ pound of sea slugs. You decide to make $\frac{3}{4}$ of the recipe. How many pounds of sea slugs will you need?

2. (a) Yes, $\frac{1}{3} \cdot \frac{2}{3}$ is the fraction of a cup of sugar you should use, since you will need $\frac{1}{3}$ of $\frac{2}{3}$ of a cup of sugar.

   (b) No, from the information given, we don't know if $\frac{1}{3}$ *of the cars that have power steering* have side mounted airbags.

(c) Yes, $\frac{1}{3}$ of $\frac{2}{3}$ of the cars at the car dealership have both power steering and side mounted airbags.

(d) No, we can't tell *how many* candies Ed ate, only what *fraction of a bag* of candies Ed ate.

(e) Yes, Ed ate $\frac{1}{3}$ of $\frac{2}{3}$ of a bag of candies.

3. See the text.

4. No, although Anthony has drawn a good representation of the problem, his reasoning is not completely right. 10 pieces are shaded, and these 10 pieces do represent $\frac{5}{7}$ of 2 rectangles. But Anthony must remember that each small piece represents $\frac{1}{7}$ *of the original object*, which we infer must have been one rectangle. So $\frac{5}{7} \cdot 2$ is the fraction of *one* rectangle that is shaded. This fraction is $\frac{10}{7}$. When drawing pictures such as Anthony's, it's a good idea to also draw 1 whole object somewhere as a reminder that this is your reference amount.

5.

$$
\begin{aligned}
\frac{2}{3} \times 6\frac{3}{4} &= \frac{2}{3} \times (6 + \frac{3}{4}) \\
&= \frac{2}{3} \times 6 + \frac{2}{3} \times \frac{3}{4} \\
&= 4 + \frac{1}{2} \\
&= 4\frac{1}{2}.
\end{aligned}
$$

The distributive property was used at the second equal sign.

Problems for Section 4.9 on Multiplying Fractions

1. Make up a story problem for

$$
\frac{1}{3} \cdot \frac{1}{4}.
$$

Use the *meaning* of fractions and the *meaning* of multiplication, and use *pictures* to determine the answer to the multiplication problem. Explain clearly.

2. Make up a story problem for

$$2 \cdot \frac{3}{5}.$$

Use the *meaning* of fractions and the *meaning* of multiplication, and use *pictures* to determine the answer to the multiplication problem. Explain clearly.

3. Make up a story problem for

$$\frac{2}{3} \cdot 5.$$

Use the *meaning* of fractions and the *meaning* of multiplication, and use *pictures* to determine the answer to the multiplication problem. Explain clearly.

4. (a) Make up a story problem for $2\frac{3}{4} \times 3\frac{1}{2}$.

   (b) Use pictures and the meaning of multiplication to solve the problem.

   (c) Use the distributive property or FOIL to calculate $2\frac{3}{4} \times 3\frac{1}{2}$ by rewriting this product as $(2 + \frac{3}{4}) \times (3 + \frac{1}{2})$.

   (d) Identify the four terms produced by the distributive property or FOIL (in part (c)) in a picture like one in part (b).

   (e) Now write the mixed numbers $2\frac{3}{4}$ and $3\frac{3}{4}$ as improper fractions and use the standard procedure for multiplying fractions to calculate $2\frac{3}{4} \times 3\frac{1}{2}$. How do you see the product of the numerators in your picture in part (b)? How do you see the product of the denominators in your picture?

5. Manda says that

$$3\frac{2}{3} \times 2\frac{1}{5} = 3 \times 2 \; + \; \frac{2}{3} \times \frac{1}{5}.$$

Explain why Manda has made a good attempt, but her answer is not correct. Explain how to *work with what Manda has already written* and modify it to get the correct answer (in other words, don't just start from scratch and show Manda how to do the problem, but rather take what she has already written, use it, and make it mathematically correct). Are any properties of arithmetic involved in this? How is this related to work (other than on fractions) that we have done in this chapter?

6. Let's suppose that the liquid in a car's radiator is 75% water and 25% antifreeze. Suppose 30% of the radiator's liquid is drainded out and replaced with pure antifreeze. Now what percent of the radiator's liquid is antifreeze? Draw pictures to help you solve this problem. Explain your answer clearly.

4.10   Powers

In Chapter 2 we studied powers of 10, which are the foundation of the decimal system. More generally, we can consider powers of numbers other than 10. When powers are multiplied, they behave in a very nicely described fashion. In the next section, when we study the standard procedure for multiplying decimals, it will be useful to know how of powers of 10 behave when they are multiplied.

Class Activity 4DD: Multiplying Powers of 10

Powers of Numbers Other Than 10

Just as 10^5 stands for five tens multiplied together:

$$10^5 = 10 \times 10 \times 10 \times 10 \times 10,$$

the expression 2^5 stands for five 2s multiplied together:

$$2^5 = 2 \times 2 \times 2 \times 2 \times 2.$$

Similarly 3^4 stands for four 3s multiplied together:

$$3^4 = 3 \times 3 \times 3 \times 3.$$

In general, if A is any real number and B is any counting number, then A^B stands for B As multiplied together.

$$A^B = \underbrace{A \times A \times \ldots \times A}_{B \text{ times}}$$

As with powers of 10, we read

$$A^B$$

as *A to the Bth power* or *A to the B*, and *B* is called the **exponent** of A^B.

If you did Class Activity 2A in Chapter 2, then you saw that powers of 10 become very large very quickly. Even powers of 2 grow very large quickly, although they start off more slowly than powers of 10:

$$
\begin{aligned}
2^1 &= 2 \\
2^2 &= 4 \\
2^3 &= 8 \\
2^4 &= 16 \\
2^5 &= 32 \\
2^6 &= 64 \\
2^7 &= 128 \\
2^8 &= 256 \\
2^9 &= 512 \\
2^{10} &= 1,024 \\
&\vdots \\
2^{20} &= 1,048,576 \\
&\vdots \\
2^{30} &= 1,073,741,824
\end{aligned}
$$

When a sequence of numbers consists of consecutive powers of one fixed number that is greater than 1, then we say that this sequence has **exponential growth**. The children's book *One Grain of Rice* [14], beautifully illustrates the exponential growth exhibited by powers of 2.

Exponential growth occurs in many natural systems. Population growth, whether of bacteria, animals, or people, can grow exponentially when there is enough food and space, and when there are no predators. Money left to grow in a bank account with a fixed interest rate grows exponentially. Similarly, debt, when it is not paid off, grows exponentially.

Exercises for Section 4.10 on Powers

1. Explain why $10^A \times 10^B = 10^{A+B}$ is always true whenever A and B are any counting numbers.

Answers to Exercises for Section 4.10 on Powers

1. The expression $10^A \times 10^B$ stands for A tens multiplied by B tens. All together, that is $A + B$ tens multiplied together. We can show this with the following equations:

$$10^A \times 10^B = \underbrace{10 \times 10 \times \ldots \times 10}_{A \text{ times}} \times \underbrace{10 \times 10 \times \ldots \times 10}_{B \text{ times}}$$
$$= \underbrace{10 \times 10 \times \ldots \times 10 \times 10 \times 10 \times \ldots \times 10}_{A+B \text{ times}}$$

Problems for Section 4.10 on Powers

1. Is 9^5 equal to $99,999$? Explain.

2. (a) Use a calculator to compute 2^{161}. Are you able to tell what the ones digit of this number is from the calculator's display? Why or why not?

   (b) Determine the ones digits of each of the numbers $2^1, 2^2, 2^3, 2^4, 2^5, \ldots, 2^{17}$. Describe the pattern in the ones digits of these numbers.

   (c) Using the pattern that you discovered in (b), predict the ones digit of 2^{161}. Explain your reasoning clearly.

3. (a) Use the meaning of exponents to show how to write the expressions in (i), (ii), and (iii) below as a *single* power of ten, i.e., in the form 10^A for some A.

     i. $(10^2)^3$
     ii. $(10^3)^3$
     iii. $(10^4)^2$

   (b) In each of (i), (ii), and (iii) above, relate the exponents in the original expression to the exponent in the answer. Describe this relationship.

(c) Explain why it is always true that $(10^A)^B = 10^{A \times B}$ whenever A and B are any natural numbers.

4. (a) Explain why the shaded portions of the four figures in Figure 4.36 represent

$$\frac{1}{2},$$

$$\frac{1}{2} + (\frac{1}{2})^2,$$

$$\frac{1}{2} + (\frac{1}{2})^2 + (\frac{1}{2})^3,$$

and

$$\frac{1}{2} + (\frac{1}{2})^2 + (\frac{1}{2})^3 + (\frac{1}{2})^4$$

of the figures respectively.

 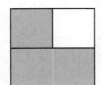

Figure 4.36: Four Figures

(b) What fraction of the figure is shaded in the 2nd, 3rd, and 4th figure? Give each answer in simplest form. If you were to continue this process, when would you get to a figure that is at least 99.9% shaded?

(c) Imagine that you could continue the above process of shading figures forever. Based on the pictures above, what would you expect the infinite sum

$$\frac{1}{2} + (\frac{1}{2})^2 + (\frac{1}{2})^3 + (\frac{1}{2})^4 + (\frac{1}{2})^5 + \dots$$

to be equal to?

(d) Here's another way to see what the infinite sum in part (c) is. Let S stand for this sum. In other words,

$$S = \frac{1}{2} + (\frac{1}{2})^2 + (\frac{1}{2})^3 + (\frac{1}{2})^4 + (\frac{1}{2})^5 + \ldots$$

Find another way to express $\frac{1}{2} + \frac{1}{2}S$ as follows. Start by writing

$$\frac{1}{2} + \frac{1}{2}S = \frac{1}{2} + \frac{1}{2}\left(\frac{1}{2} + (\frac{1}{2})^2 + (\frac{1}{2})^3 + (\frac{1}{2})^4 + (\frac{1}{2})^5 + \ldots\right).$$

Then apply the distributive property (assume that a version of the distributive property applies to infinite sums). Use this to explain why

$$\frac{1}{2} + \frac{1}{2}S = S. \qquad (4.12)$$

Then use Equation 4.12 to solve for S, i.e., to find a number S that makes Equation 4.12 true. Summarize what you did in this part and what conclusion this leads you to.

4.11   Multiplying Decimals

To multiply (finite) decimals, the standard procedure is to first multiply the numbers obtained by deleting the decimal points, and then to place the decimal point in the answer according to a certain rule. Why is this procedure valid? The class activities and exercises in this section will help you answer this question.

Once we know how to multiply fractions, we really don't even need to know a rule for multiplying finite decimals. This is because finite decimals are special kinds of fractions: ones that can be written with a denominator that is a power of 10. For example,

$$13.48 = \frac{1348}{100}, \quad 67.9 = \frac{679}{10}.$$

From this point of view, multiplying finite decimals is just a special case of multiplying fractions:

$$13.48 \times 67.9 \quad = \quad \frac{1348}{100} \times \frac{679}{10}$$

$$= \frac{1348 \times 679}{100 \times 10}$$

$$= \frac{915292}{1000}$$

$$= 915.292$$

What is the standard procedure for multiplying decimals? To multiply

$$
\begin{array}{r}
1.36 \\
\times 2.7 \\
\hline
\end{array}
$$

you first multiply ignoring decimal points:

$$
\begin{array}{r}
136 \\
\times 27 \\
\hline
3672
\end{array}
$$

Then you add the number of digits to the right of the decimal points in your two original numbers (in this example there are 2 digits in 1.36 and another 1 digit in 2.7, for a total of $1 + 2 = 3$ digits) and you put the decimal point that many places from the end in your answer (in this example, the decimal point goes 3 digits from the end: 3.672).

Class Activity 4EE: Where Does The Decimal Point Go When We Multiply?

Exercises for Section 4.11 on Multiplying Decimals

1. Explain why the decimal point should go 4 places from the end in
$$
\begin{array}{r}
17.64 \\
\times 8.39 \\
\hline
\end{array}
$$
by comparing this product to
$$
\begin{array}{r}
1764 \\
\times 839 \\
\hline
\end{array}
$$

2. Suppose you multiply a decimal that has 5 digits to the right of its decimal point by a decimal that has 2 digits to the right of its decimal point. Explain why you put the decimal point $5 + 2$ places from the end of the product computed by ignoring the decimal points. *Work with powers of 10 to explain this.*

3. Use the meaning of multiplication and a picture to explain why 2.4×1.6 has an entry in the hundredths place, but no entries in any smaller places (thousandths, ten thousandths, etc.). Relate this to the rule for placing the decimal point in 2.4×1.6.

4. Use expanded forms and properties of arithmetic or FOIL to explain why $.31 \times .024$ has digits down to the hundred-thousandths place $\left(\frac{1}{100000}\right)$, but not in any smaller places. Relate this to the rule for placing the decimal point in $.31 \times .024$.

5. $1.35 \times 7.2 = 9.72$, but shouldn't the answer have 3 digits to the right of its decimal point? What's the problem?

Answers to Exercises for Section 4.11 on Multiplying Decimals

1. Since
$$1764 = 100 \times 17.64 = 10^2 \times 17.64$$
and
$$839 = 100 \times 8.39 = 10^2 \times 8.39$$
therefore 1764×839 is 10^2 times 10^2 times 17.64×8.39. So to get 17.64×8.39 from 1764×839, you have to *divide* 1764×839 by 10^2 and then divide again by 10^2. This means you have to divide 1764×839 by 2 tens, and then by another 2 tens, which means moving the decimal point 4 places to the *left*. In equations:

$$
\begin{aligned}
1764 \times 839 &= 10^2 \times 17.64 \times 10^2 \times 8.39 \\
&= 10^2 \times 10^2 \times 17.64 \times 8.39
\end{aligned}
$$

This means that

$$17.64 \times 8.39 = (1764 \times 839 \div 10^2) \div 10^2$$

which means moving the decimal point $2 + 2 = 4$ places to the left.

2. Let's work with a particular example to illustrate, say 1.23456×1.23. Since
$$123456 = 10^5 \times 1.23456$$

and

$$123 = 10^2 \times 1.23,$$

therefore 123456×123 is $10^5 \times 10^2$ times 1.23456×1.23. So to get 1.23456×1.23 from 123456×123 you have to divide 123456×123 by 10^5 and then by 10^2. This means dividing by 5 tens, and then dividing by another 2 tens. Altogether, that's dividing by $5 + 2 = 7$ tens. This is why you move the decimal point $5 + 2 = 7$ places to the left in the answer to 123456×123 in order to get the correct answer to 1.23456×1.23. Notice that the explanation works the same way for any other numbers with 5 and 2 digits to the right of their decimal points.

3. 2.4×1.6 means the total number of objects in 2.4 groups when there are 1.6 objects in each group. Starting with a square, Figure 4.37 shows 1.6 of the square, then 2 groups of 1.6 of the square and finally 2.4 groups of the 1.6 of the square. In the 2.4 groups of the 1.6 of the square there are 2 whole squares, a bunch of "strips," each of which is $\frac{1}{10}$ of a square, and a bunch of small squares, each of which is $\frac{1}{100}$ of a square. Therefore, visibly, 2.4×1.6 has entries in the tenths and hundredths places, but not in any smaller places. This agrees with the rule of placing the decimal point $1 + 1 = 2$ places from the end in 24×16 in order to get the answer to 2.4×1.6, because this will produce entries in the tenths and hundredths places, but not in any smaller places.

4.

$$
\begin{aligned}
.31 \times .024 &= \left(3 \cdot \frac{1}{10} + 1 \cdot \frac{1}{10^2}\right) \times \left(2 \cdot \frac{1}{10^2} + 4 \cdot \frac{1}{10^3}\right) \\
&= 3 \cdot 2 \cdot \frac{1}{10} \cdot \frac{1}{10^2} + 3 \cdot 4 \cdot \frac{1}{10} \cdot \frac{1}{10^3} + 1 \cdot 2 \cdot \frac{1}{10^2} \cdot \frac{1}{10^2} + 1 \cdot 4 \cdot \frac{1}{10^2} \cdot \frac{1}{10^3} \\
&= 6 \frac{1}{10^3} + 12 \cdot \frac{1}{10^4} + 2 \cdot \frac{1}{10^4} + 4 \cdot \frac{1}{10^5}
\end{aligned}
$$

We could keep on going and regroup to get the final answer, but this already demonstrates that there are entries down to the $\frac{1}{10^5}$ place, but no lower. This is consistent with the rule for placing the decimal point $2 + 3 = 5$ places from the end of 31×24, because that will produce entries down to the $\frac{1}{10^5}$ place but no lower.

Thinking more generally about what will happen with other numbers, notice that the smallest place value in the product will always occur

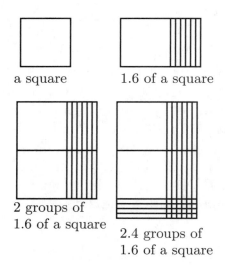

Figure 4.37: Showing 2.4×1.6

when the two smallest place values are multiplied. If you multiply $\frac{1}{10^A}$ by $\frac{1}{10^B}$ the result is $\frac{1}{10^{A+B}}$, so again this shows why we add the number of places behind the decimal points.

5. The rule says to find the product without the decimal points, i.e., 135×72 and then move the decimal point *in that answer* $2 + 1 = 3$ places to the left. But

$$135 \times 72 = 9720$$

ends in a zero. That's why that third digit doesn't show up in the answer.

Problems for Section 4.11 on Multiplying Decimals

1. Use the meaning of multiplication and use a picture to help you explain why 1.7×3.1 has entries down to the hundredths place, but not in any place of smaller value. Be sure to explain clearly how you are interpreting your picture.

2. Leah is working on the multiplication problem 3.57×41.809. She multiplies $357 \times 41,809$ longhand and gets 14925813. But Leah can't re-

member the rule about where to put the decimal point in 14925813 to get the correct answer to 3.57×41.809.

(a) How can Leah use *reasoning about the sizes of the numbers* to figure out where to put the decimal point? (*Do not* just tell Leah about the standard procedure for where to put the decimal point.)

(b) Write 41.809 and 3.57 as *improper fractions* whose denominators are *powers of 10*. Then use *fraction multiplication* to explain where the decimal point should be placed in 14925813 to get the correct answer to 3.57×41.809. *Show how to use the powers of 10 in the denominators to explain why we add the number of digits behind the decimal points in 41.809 and 3.57.*

4.12   Multiplying Negative Numbers

So far we have examined multiplication only for numbers that are not negative. What is the meaning of multiplication for negative numbers? In this section we will examine several different ways to make sense of multiplication when negative numbers are involved.

What should 3×-2 mean? We know that 3×2 stands for the total number of objects in 3 groups if there are 2 objects in each group. If we try to use the same interpretation for 3×-2, then we say:

3×-2 is the total number of objects in 3 groups if there are -2 objects in each group.

But what does this mean? One way to interpret this sensibly is to think of

"-2 objects in each group"

as meaning

"each group *owes* 2 objects."

For example, if Lakeisha, Mary, and Jayna each owe $2.00, then we can think of each of the girls as *having* -2 dollars, and all together, the 3 girls have 3×-2 dollars. Notice that with this interpretation, it makes sense to say

$$3 \times -2 = -6$$

because all together, the 3 girls collectively owe 6 dollars—this is represented by the -6. In general, if A and B are positive numbers (or zero) then we can define $A \times -B$ to be $-(A \times B)$, i.e.,

$$A \times -B = -(A \times B) \tag{4.13}$$

and this definition is consistent with the interpretation of negative numbers as amounts *owed*. Equation 4.13 gives us the familiar rule:

A positive times a negative is negative.

What about a negative number times a positive number, such as

$$-2 \times 3?$$

We could perhaps start talking about "groups owed," but notice that *if we expect the commutative property of multiplication to hold for both positive and negative numbers*, then

$$-2 \times 3 = 3 \times -2,$$

which must be equal to -6 according to Equation 4.13 and the discussion above. In general, if A and B are any positive numbers, then in order to have a number system that behaves in a manner consistent with the behavior of positive numbers, we must define $-A \times B$ to be $-(A \times B)$, i.e.,

$$-A \times B = -(A \times B) \tag{4.14}$$

This equation gives us the familiar rule:

A negative times a positive is negative.

Finally, what about a negative number times a negative number, such as

$$-4 \times -5?$$

It's hard to see how to make sense of something like "the total number of objects in 4 owed groups with each group owing 5 objects." Instead, we can once again appeal to the idea of expecting the familiar properties of arithmetic to hold for both positive and negative numbers. So *if we expect the distributive property to hold for both positive and negative numbers*, then

$$
\begin{aligned}
-4 \times -5 + 4 \times -5 &= (-4 + 4) \times -5 \\
&= 0 \times -5 \\
&= 0.
\end{aligned}
$$

This means that -4×-5 and 4×-5 add to zero. Since $4 \times -5 = -20$ (by Equation 4.13), therefore -4×-5 and -20 add to zero. But 20 is the only number which, when added to -20, yields zero. Therefore -4×-5 must equal 20. In general, if A and B are any positive numbers, then in order to have a number system that behaves in a manner consistent with the behavior of positive numbers, we must define $-A \times -B$ to be $A \times B$, i.e.,

$$
-A \times -B = A \times B. \tag{4.15}
$$

This gives us the familiar rule:

A negative times a negative is positive.

Class Activity 4FF: Patterns With Multiplication and Negative Numbers

Exercises for Section 4.12 on Multiplying Negative Numbers

1. Explain why it makes sense to say that $3 \times -2 = -6$.

2. Explain why it makes sense to say that $-2 \times 3 = 6$.

3. Explain why it makes sense to say that $-4 \times -5 = 20$.

Answers to Exercises for Section 4.12 on Multiplying Negative Numbers

1. See text.

2. See text.

3. See text.

Problems for Section 4.12 on Multiplying Negative Numbers

1. In ordinary language, the term "multiply" means "make larger," as in:

   Go forth and multiply.

   In mathematics, does multiplying always make larger? In other words, if you start with a number N and multiply it by another number M, is the resulting product $M \times N$ larger than N?

   (a) Investigate this question by working out a number of examples. M and N can be *any* kinds of numbers: they can be fractions, integers, or decimals, they can be positive or negative.

   (b) Formulate a general answer to the following question: given a number N, for which numbers M is $M \times N$ greater than N? (Your answer should involve cases, depending on the nature of N.)

4.13   Scientific Notation

Many scientific applications require the use of very large or very small numbers: distances between stars are huge, whereas molecular distances are tiny. These kinds numbers can be cumbersome to write in ordinary decimal notation. Therefore, a special notation, called *scientific notation* is often used to write such numbers. Scientific notation involves multiplying by powers of 10.

Use a calculator to multiply

$$123456789 \times 987654321.$$

How is the answer displayed? The answer is probably displayed in the form:

$$1.2193263 \quad 17$$

or

$$1.2193263 \ \text{E} \ 17$$

Both of these displays mean:

$$1.2193263 \times 10^{17}$$

which is in *scientific notation.*

A number is in **scientific notation** if it is written as a decimal which has exactly one nonzero digit to the left of the decimal point, times a power of ten. So a number is in scientific notation if it is written in the form

$$\#.\#\#\#\#\# \times 10^{\#},$$

where the # to the left of the decimal point is not zero, and where any number of digits can be displayed to the right of the decimal point. So neither

$$12.193263 \times 10^{16}$$

nor

$$121.93263 \times 10^{15}$$

is in scientific notation, but

$$7.39 \times 10^{47}$$

is in scientific notation.

Other than 0, every real number can be expressed in scientific notation. To put a number in scientific notation, think about how multiplication by powers of 10 works. For example, how do we write

$$847,930,000$$

in scientific notation? We need to find an exponent that will make the following equation true:

$$847,930,000 = 8.4793 \times 10^{?}.$$

The decimal point in 8.4793 must be moved 8 places to the right to get 847,930,000, therefore we should multiply 8.4793 by 10^{8}. In other words,

$$847,930,000 = 8.4793 \times 10^{8}.$$

How do we write

$$.0000345$$

in scientific notation? We need to find an exponent that will make the following equation true:

$$.0000345 = 3.45 \times 10^{?}.$$

The decimal point in 3.45 must be moved 5 places to the *left* to get .0000345, therefore we must multiply 3.45 by

$$.00001 = \frac{1}{10^5} = 10^{-5}$$

in order to get .0000345. So

$$.0000345 = 3.45 \times 10^{-5}.$$

Here are a few more examples:

Ordinary decimal notation:	Scientific notation:
12	1.2×10^1
123	1.23×10^2
1234	1.234×10^3
12345	1.2345×10^4
1234.5	1.2345×10^3
123.45	1.2345×10^2
12.345	1.2345×10^1
1.2345	1.2345 or 1.2345×10^0
.12345	1.2345×10^{-1}
.012345	1.2345×10^{-2}
.0012345	1.2345×10^{-3}

Although scientific notation is mainly used in scientific settings, it is common to use a variation of scientific notation when discussing large numbers in more common situations, such as when budgets or populations are concerned. For example, an amount of money may be described as

$3.5 billion,

which means

$$\$ \, 3.5 \times (\text{one billion})$$

or

$$\$ \, 3.5 \times 1,000,000,000.$$

Other ways to express $3.5 billion are:

$$\$ \, 3.5 \times 10^9,$$

or just

$$\$3,500,000,000.$$

The form

$$\$3.5 \text{ billion}$$

is probably more quickly and easily grasped by most people than any of the other forms.

By the way, what is a big number, or a small number? Children sometimes ask questions like:

Is 100 a big number? What about 1000, is that a big number?

The answer is: "it depends". 1000 grains of sand is not much sand at all, but 10 pages of homework is a lot!

Class Activity 4GG: Scientific Notation Versus Ordinary Decimal Notation

Class Activity 4HH: How Many Digits are in a Product of Counting Numbers?

Class Activity 4II: Explaining the Pattern in the Number of Digits in Products

Exercises for Section 4.13 on Scientific Notation

1. Write the following numbers in scientific notation.

   (a) $153,293,043,922$
   (b) $.00000321$
   (c) $(2.398 \times 10^{15}) \times (3.52 \times 10^9)$
   (d) $(5.9 \times 10^{15}) \times (8.3 \times 10^9)$

2. The following populations of states are from the 2000 census (see www.census.gov). In each case, describe the population, rounded to the nearest hundred thousand, in a way that would be understandable to most people.

   (a) Georgia: $8,186,453$

    (b) Florida: $15,982,378$

    (c) California: $33,871,648$

    (d) Rhode Island: $1,048,319$

    (e) New York: $18,976,457$

3. The astronomical unit (AU) is used to measure distances. One AU is the average distance from the earth to the sun, which is 92,955,630 miles. We are 2 billion AU from the center of the Milky Way galaxy (our galaxy). How many *miles* are we from the center of the Milky Way galaxy? Give your answer in scientific notation. Explain how the meaning of multiplication applies to this problem.

4. According to scientific theories, the solar system formed between 5 and 6 billion years ago. Light travels 186,282 miles per second. How far has the light from the forming solar system traveled in 5.5 billion years? Give your answer in scientific notation.

5. Sam uses a calculator to multiply $666,666 \times 7,777,777$. The calculator's answer is displayed as follows:

$$5.1851794 \; E \; 12$$

So Sam writes:

$$666,666 \times 7,777,777 = 5,185,179,400,000.$$

Is Sam's answer correct or not? If not, why not?

4.13.1   Answers to Exercises for Section 4.13 on Scientific Notation

1.   (a) $1.53293043922 \times 10^{11}$

    (b) 3.21×10^{-6}

    (c) 8.44096×10^{24}

    (d) 4.897×10^{25}

2.   (a) Georgia: 8.1 million

    (b) Florida: 16.0 million

(c) California: 33.9 million

(d) Rhode Island: 1.0 million

(e) New York: 19.0 million

3. Since we are 2 billion AU from the center of the galaxy and each AU is
    92,955,630 miles, the number of miles from the earth to the center of
    the galaxy is the total number of objects in 2 billion groups (each AU is
    a group) when there are 92,955,630 objects in each group (each object
    is 1 mile). Therefore, according to the meaning of multiplication, we
    are

$$
\begin{aligned}
\text{2 billion} \times 92,955,630 &= 2 \times 10^9 \times 92,955,630 \\
&= 185,911,260 \times 10^9 \\
&= 1.859 \times 10^8 \times 10^9 \\
&= 1.859 \times 10^{17}
\end{aligned}
$$

    miles from the center of the galaxy.

4. 3.23×10^{22} miles

5. No, Sam's answer is not correct. When the calculator displays the
    answer to $666,666 \times 7,777,777$ as

$$5.1851794 \; E \; 12,$$

    this stands for

$$5.1851794 \times 10^{12}.$$

    But the calculator is forced to round its answer because it can only
    display so many digits on its screen. Therefore, although it is true that

$$5.1851794 \times 10^{12} = 5,185,179,400,000,$$

    this is not the exact answer to $666,666 \times 7,777,777$. Instead, it is
    the answer to $666,666 \times 7,777,777$ *rounded to the nearest hundred-
    thousand.*

Problems for Section 4.13 on Scientific Notation

1. Tanya says that the ones digit of 2^{59} is a 5 because her calculator's display for 2^{59} reads:

$$5.76460752303E17,$$

   and there is a 5 in the ones place. Is Tanya right? Why or why not?

2. Suppose that a laboratory has one gram of a radioactive substance that has a half-life of 100 years. "A half-life of 100 years" means that no matter what amount of the radioactive substance one starts with, after 100 years, only half of it will be left. So, after the first 100 years, only half a gram would be left, and after another 100 years, only one quarter of a gram would be left.

   (a) How many hundreds of years will it take until there is less than one hundred millionth of a gram left of the laboratory's radioactive substance? (Give a whole number of hundreds of years.)

   (b) How many hundreds of years will it take until there is less than one billionth of a gram left of your radioactive substance? (Give a whole number of hundreds of years.)

3. Light is known to travel at a speed of about $300,000$ kilometers per second.

   (a) How far does light travel in one day? Give your answer both in scientific notation and in ordinary decimal notation. Explain your work.

   (b) How far does light travel in one year? Give your answer both in scientific notation and in ordinary decimal notation, rounded to the nearest hundred billion kilometers. Explain your work. The distance that light travels in one year is called a *light year*.

4. Is 2×10^7 equal to 2^7? Is 1×10^9 equal to 1^9? If not, explain the distinctions between the expressions.

5. Let's say that you want to write the answer to $179234652 \times 437481694$ as a whole number in ordinary decimal notation, showing all its digits. Find a way to use a calculator to help you do this in an efficient way. *Do not just multiply longhand.* Explain your technique and explain why it works.

6. According to the Department of the Treasury, the gross federal debt in 2001 was approximately $5.8 trillion (see www.ustreas.gov/opc/). Predict the federal debt in 2004, assuming that the federal debt goes up by 5% every year. (To say that the debt goes up by 5% each year means that each year's debt is 5% more than *the previous year's debt*.) Give your answer rounded to the nearest hundred-billion dollars.

7. Suppose you owe $1,000 on your credit card and that at the end of each month an additional 1.4708% of the amount you owe is added on to the amount you owe. (This is what would happen if your credit card charged an annual percentage rate of 17.6496%.) Let's assume that you do not pay off any of this debt or the interest that is added to it. Let's also assume that you don't add on any more debt (other than the interest that you are charged).

   (a) How much will you owe after 6 months? after one year? after two years?

   (b) Calculate $(1.014708)^{48}$ and notice that this number is just a little bigger than 2. Based on this, what will happen to your debt every 4 years? Explain.

   (c) Suppose you didn't pay off your debt for 40 years. Using part (b), determine approximately how much you owe?

Chapter 5

Division

In this chapter we will study division of all different kinds of numbers: whole numbers, rational numbers, real numbers, and integers. As with multiplication, it is easy to take the concept of division for granted and to dismiss it as a trivial. However, even though it is easy for adults to carry out the *procedure* of dividing, the underlying *concept* of division is much more subtle. In this chapter, we will study the concept of division, especially the different ways of interpreting the meaning of division. We will see that the *meaning* of division gives rise to the familiar procedures that we use when we divide whole numbers, fractions, and decimals. We will also study the link between fractions and division.

5.1   The Meaning of Division

In this section we will study the meaning of division. There are two main ways to interpret the meaning of division; both are useful and both are natural. Just as every subtraction problem can be re-written as an addition problem, so too every division problem can be re-written as a multiplication problem. We will use this to explain why a division problem has the same answer, regardless of which of the two main interpretations of division we use. Finally, each of the two main ways of interpreting the meaning of division is subject to further interpretation, depending on whether one does or does not allow for a remainder.

There are three standard ways to denote division:

$$A \div B \quad A/B \text{ and } B\overline{)A}$$

All three are read "A divided by B." In a division problem $A \div B$, the result is called the **quotient**, the number A is called the **dividend**, and the number B is called the **divisor**.

The Two Interpretations of Division

A simple example will show that there are two distinct ways to interpret the meaning of division. How would you draw a diagram to solve $8 \div 2$? You might draw something like Figure 5.1. Or you might draw a diagram like Figure 5.2.

In the first diagram, $8 \div 2$ is represented by 8 objects divided into *groups of* 2. With this interpretation we're thinking of $8 \div 2 = ?$ as asking how

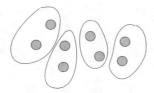

Figure 5.1: One View of $8 \div 2$

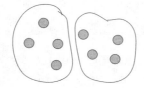

Figure 5.2: Another View of $8 \div 2$

many groups there will be if 8 objects are divided into groups with 2 objects in each group. Whereas in the second diagram, $8 \div 2$ is represented by 8 objects divided into 2 *groups*. With the second interpretation, we're thinking of $8 \div 2 =?$ as asking how many objects will be in each group if 8 objects are divided equally among 2 groups.

The *How Many Groups?* Interpretation

If A and B are non-negative numbers, and B is not 0, then according to the **how many groups?** interpretation of division:

> $A \div B$ means the number of groups when A objects are divided
> into groups with B objects in each group.

In other words, with this interpretation, $A \div B$ means "the number of Bs that are in A." This interpretation of division is sometimes called the **measurement model of division** or the **subtractive model of division**.

With this interpretation of division, $18 \div 6 = 3$ because if you have 18 objects and you divide these into groups of 6, then there will be 3 groups, as shown in Figure 5.3.

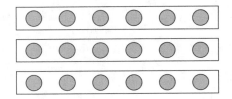

Figure 5.3: $18 \div 6 = 3$

The *How Many in Each Group?* Interpretation

If A and B are non-negative numbers and B is not 0, then according to the **how many in each group?** interpretation of division:

> $A \div B$ means the number of objects in each group when A objects are divided equally among B groups.

This interpretation of division is sometimes called the **partitive model of division** or the **sharing model of division**.

With this interpretation of division we conclude that $18 \div 6 = 3$ because if you divide 18 objects equally among 6 groups, then there will be 3 objects in each group, as seen in Figure 5.4.

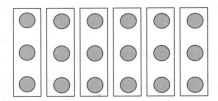

Figure 5.4: $18 \div 6 = 3$

Why do we get the same answer to division problems, regardless of which interpretation of division we use? We will answer this by relating division to multiplication.

Relating Division and Multiplication

Consider the division problem

$$18 \div 6 = ?$$

According to the *how many groups?* interpretation of division, $18 \div 6$ is the number of groups of 6 that make 18. Therefore, according to the meaning of multiplication, that number of groups times 6 is equal to 18. In other words, with this interpretation of division,

$$18 \div 6 = ?$$

means the same as

$$? \times 6 = 18.$$

In the same way, every division problem is equivalent to a multiplication problem. With the *how many groups?* interpretation of division, the problem

$$A \div B = ?$$

is equivalent to the problem

$$? \times B = A.$$

On the other hand, according to the *how many in each group?* interpretation of division, $18 \div 6$ is the number of objects for which 6 groups of that many objects makes a total of 18 objects. Therefore, according to the meaning of multiplication, 6 times that number of objects is equal to 18. In other words, with this interpretation of division,

$$18 \div 6 = ?$$

means the same as

$$6 \times ? = 18.$$

In the same way, every division problem is equivalent to a multiplication problem. With the *how many in each group?* interpretation of division, the problem

$$A \div B = ?$$

is equivalent to the problem

$$B \times ? = A.$$

Now we can explain why a division problem will have the same answer, regardless of whether the *how many groups?* or the *how many in each group?* interpretation of division is used. The key lies in the commutative property of multiplication. According to the *how many groups?* interpretation, to solve

$$A \div B = ?$$

is equivalent to solving

$$? \times B = A.$$

But according to the commutative property of multiplication,

$$? \times B = B \times ?.$$

Therefore solving

$$? \times B = A$$

is equivalent to solving

$$B \times ? = A,$$

which is equivalent to solving

$$A \div B = ?$$

with the *how many in each group?* interpretation of division. Therefore, we will always get the same answer to a division problem regardless of which way we interpret division.

Answers With or Without Remainders

Consider the division problem $23 \div 4$. Depending on the context, any one of the following three answers could be most appropriate:

$$23 \div 4 = 5.75$$
$$23 \div 4 = 5\frac{3}{4}$$
$$23 \div 4 = 5, \text{ remainder } 3$$

The first two answers fit with division as we have been interpreting it so far. For example, to divide \$23 equally among 4 people, each person should get $23 \div 4 = \$5.75$. Or, if you have 23 cups of flour, and a batch of cookies requires 4 cups of flour, then you can make $23 \div 4 = 5\frac{3}{4}$ batches of cookies (assuming that it is possible to make $\frac{3}{4}$ of a batch of cookies). But if you have 23 pencils to divide equally among 4 children, then it doesn't make sense to give each child 5.75 or $5\frac{3}{4}$ pencils. Instead, it is best to give each child 5 pencils and keep the remaining 3 pencils in reserve. How does this answer, 5, remainder 3, fit with division as we have defined it?

In order to get the third answer, 5, remainder 3, we need to interpret division in a slightly different manner than we have so far. For each of the two main interpretations of division, there is an alternate formulation, which allows for a remainder. In these alternate formulations, we seek a quotient that is a *whole number*.

If A and B are whole numbers, and B is not zero, then

$A \div B$ is the largest *whole number* of objects in each group when A objects are divided equally among B groups. The **remainder** is the number of objects left over, i.e., that can't be placed in a group. (*How many in each group?* interpretation, with remainder)

$A \div B$ is the largest *whole number* of groups that can be made when A objects are divided into groups with B objects in each group. The **remainder** is the number of objects left over, i.e., that can't be placed in a group. (*How many groups?* interpretation, with remainder)

5.1.1   Division With Negative Numbers

Above, division was defined for non-negative numbers. How can we make sense of division problems like

$$-18 \div 3 = ?$$

or

$$18 \div -3 = ?$$

We could perhaps talk about dividing 18 owed objects among 3 groups, but this seems strange. A better way is to re-write a division problem with negative numbers as a multiplication problem. For example, the problem

$$-18 \div 3 = ?$$

is equivalent to

$$? \times 3 = -18$$

(if we use the *how many groups?* interpretation). Since $-6 \times 3 = -18$, and since no other number times 3 is -18, therefore

$$-18 \div 3 = -6.$$

By considering how multiplication works when negative numbers are involved, we can see that the following rules apply to division problems involving negative numbers:

$$\begin{aligned} \text{negative} \div \text{positive} &= \text{negative} \\ \text{positive} \div \text{negative} &= \text{negative} \\ \text{negative} \div \text{negative} &= \text{positive} \end{aligned}$$

Class Activity 5A: The Two Interpretations of Division

Class Activity 5B: Remainder, Mixed Number, or Decimal Answer to Division Problem: Which is Most Appropriate?

Class Activity 5C: Why Can't We Divide by Zero?

Class Activity 5D: Can We Use Properties Of Arithmetic To Divide?

Exercises For Section 5.1 on the Meaning of Division

1. For each of the following story problems, write the corresponding division problem and decide which interpretation of division is involved (the *how many groups?* or the *how many in each group?*, with or without remainder).

   (a) There are 5237 tennis balls that are to be put into packages of 3. How many packages of balls can be made and how many will be left over?

   (b) If 50 fluid ounces of mouthwash costs $4.29, then what is the price per fluid ounce of this mouthwash?

   (c) If 50 fluid ounces of mouthwash costs $4.29, then how much of this mouthwash is worth $1?

(d) If you have a full 50 fluid ounce bottle of mouthwash and you use 3 fluid ounces per day, then how many days will this mouthwash last?

(e) 1 yard is 3 feet. How long is an 80 foot long stretch of sidewalk in yards?

(f) 1 gallon is 16 cups. How many gallons are 65 cups of lemonade?

(g) If you drive 316 miles at a constant speed and it takes you 5 hours, then how fast did you go?

2. Make up two story problems for $300 \div 12$, one for each of the two interpretations of division.

3. Josh says that $0 \div 5$ doesn't make sense because if you have nothing to divide among 5 groups, then you won't be able to put anything in the 5 groups. Which interpretation of division is Josh using? What does Josh's reasoning actually say about $0 \div 5$?

4. Roland says:

> I have 23 pencils to give out to my students, but I have no students. How many pencils should each student get? There's no possible answer because I can't give out 23 pencils to my students if I have no students!

Write a division problem that corresponds to what Roland said. Which interpretation of division does this use? What does Roland's reasoning say about the answer to this division problem?

5. Katie says:

> I have no candies to give out and I'm going to give each of my friends 0 candies. How many friends do I have? I could have 10 friends, or 50 friends, or 0 friends—there's no way to tell!

Write a division problem that corresponds to what Katie said. Which interpretation of division does this use? What does Katie's reasoning say about the answer to this division problem?

6. Use the *how many groups?* interpretation of division to explain why $1 \div 0$ is not defined.

7. Explain why $0 \div 0$ is not defined by rewriting $0 \div 0 = ?$ as a multiplication problem.

8. Marina says that since $300 \div 50 = 6$, that means you can put 300 marbles into 6 groups with 50 marbles in each group. Marina thinks that this should help her calculate $300 \div 55$. At first she tries $300 \div 50 + 300 \div 5$, but that answer seems too big. Is there some other way to use the fact that $300 \div 50 = 6$ in order to calculate $300 \div 55$?

9. According to one study, the average American produces 4.21 pounds of solid waste per day. The population of the U.S. is about 275 million.

   (a) How many pounds of solid waste does the population of the U.S. produce in one year?

   (b) If fifty pounds of solid waste take up one cubic foot, then how many cubic feet of solid waste does the U.S. population produce per year?

   (c) If the solid waste produced by the U.S. population in a year was formed into a large cube, what would the length, width and depth of this cube be?

10. Which is a better buy: 67 cents for 58 ounces or 17 cents for 15 ounces?

11. If March 5th is a Wednesday, then why do you know right away that March 12th, 19th, 26th and April 2nd are Wednesdays? Explain this using mathematics and knowledge of our calendar system.

12. Suppose that today is Friday. What day of the week will it be 36 days from today? What day of the week will it be 52 days from today? What about 74 days from today? Explain how division with remainder is relevant to these questions.

Answers to Exercises for Section 5.1 on the Meaning of Division

1. (a) $5237 \div 3$. This uses the *how many groups?*, with remainder interpretation.

(b) $\$4.29 \div 50$. This uses the *how many in each group?* interpretation (without remainder). Each ounce of mouthwash represents a group. We want to divide $\$4.29$ equally among these groups.

(c) $50 \div 4.29$. This uses the *how many in each group?* interpretation (without remainder). Each dollar represents a group. We want to divide 50 ounces up among the 4.29 groups.

(d) $50 \div 3$. This uses the *how many groups?* interpretation, with remainder. Each 3 fluid ounces is a group. We want to know how many of these groups are in 50 fluid ounces. There will be 2 fluid ounces left over.

(e) $80 \div 3$. This uses the *how many groups?* interpretation, with or without remainder. Each 3 feet is a group (a yard). We want to know how many of these groups are in 80 feet. We can give the answer as 26 yards, 2 feet (with remainder), or as 26.7 yards (without remainder).

(f) $65 \div 16$. This uses the *how many groups?* interpretation, with or without remainder. Each 16 cups is a group (a gallon). We want to know how many of these groups are in 65 cups. We can give the answer as 4 gallons and 1 cup (with remainder) or as 4.0625 gallons (without remainder). The former answer is probably more useful than the latter.

(g) $316 \div 5 = 63.2$ miles per hour. This uses the *how many in each group?* interpretation, without remainder. Divide the 316 miles equally among the 5 hours. Each hour represents a group. In each hour, you drove $316 \div 5$ miles. This means your speed was 63.2 miles per hour.

2. A good example for the *how many groups?* interpretation: "how many feet are in 300 inches?" Because each foot is 12 inches, this problem can be interpreted as "how many 12s are in 300?". An example for the *how many in each group?* interpretation is: "300 snozzcumbers will be divided equally among 12 hungry boys. How many snozzcumbers does each hungry boy get?"

3. Josh is using the *how many in each group?* interpretation of division. Josh's statement can be reinterpreted as this: if you have 0 objects and

you divide them equally among 5 groups, then there will be 0 objects in each group. Therefore $0 \div 5 = 0$.

4. The division problem that corresponds to what Roland said is this: $23 \div 0 = ?$. Roland is using the *how many in each group?* interpretation of division. Each group is represented by a student. Roland wants to divide 23 objects equally among 0 groups. But as Roland says, this is impossible to do. Therefore $23 \div 0$ is undefined.

5. The division problem that corresponds to what Kaie said is this: $0 \div 0 = ?$. Katie is using the *how many groups?* interpretation of division. Each group is represented by a friend. There are 0 objects to be distributed equally, with 0 objects in each group. But from this information, there is no way to determine how many groups there are: there could be *any* number of groups—50, or 100, or 1000. So the reason that $0 \div 0$ is undefined is because there isn't *one unique answer*, in fact *any* number could be considered equal to $0 \div 0$.

6. With the *how many groups?* interpretation, $1 \div 0$ means the number of groups when 1 object is divided into groups with 0 objects in each group. But if you put 0 objects in each group, then there's no way to distribute the 1 object—it can never be distributed among the groups, no matter how many groups there are. Therefore $1 \div 0$ is not defined.

7. Any division problem $A \div B = ?$ can be rewritten in terms of multiplication, namely as either $? \times B = A$ or as $B \times ? = A$. Therefore $0 \div 0 = ?$ means the same as $? \times 0 = 0$ or $0 \times ? = 0$. But *any number* times 0 is zero, so the ? can stand for *any number*. Since there isn't one unique answer to $0 \div 0 = ?$ we say that $0 \div 0$ is undefined.

8. Marina is right that $300 \div 50 + 300 \div 5$ won't work—it divides the 300 marbles into groups *twice*, so it gives way too many groups. Also, if you get 6 groups when you put 50 marbles in each group, then you must get *fewer groups* when you put *more marbles* in each group. If you think about removing one of those 6 groups of marbles, then you could take the 50 marbles in that group and from them, put 5 marbles in each of the remaining 5 groups. Then you'd have 5 groups of 55 marbles, and you'd have 25 marbles left over. Therefore $300 \div 55 = 5$, remainder 25.

9. (a) 4.23×10^{11} pounds.

   (b) 8.45×10^9 cubic feet

   (c) The cube would be about 2,037 feet long, wide and deep, since $2,037 \times 2,037 \times 2,037 = 8.45 \times 10^9$. (You can calculate the 2,037 by taking the cube root of 8,450,000,000 on your calculator, or by taking the cube root of 8.45 on your calculator and realizing you'll have to multiply it by 10^3. Or you could calculate the 2,037 by guessing and checking.)

10. 17 cents for 15 ounces is a slightly better buy because the price for one ounce is $17 \div 15 = 1.133$ cents, whereas when you pay 67 cents for 58 ounces the price for one ounce is $67 \div 58 = 1.155$ cents.

11. Every seven days after Wednesday is another Wednesday. March 12th, 19th, 26th, and April 2nd are 7, 14, 21, and 28 days after March 5th, and 7, 14, 21, and 28 are multiples of 7, so these days are all Wednesdays too.

12. Every seven days after a Friday is another Friday. So 35 days after today (assumed to be a Friday) is another Friday. Therefore 36 days from today is a Saturday, because it is 1 day after a Friday. Notice that we really only needed to find the remainder of 36 when divided by 7 in order to determine the answer. $52 \div 7 = 7$, remainder 3, so 52 days from today will be 3 days after a Friday (because the 7 groups of 7 get us to another Friday). 3 days after a Friday is a Monday. $74 \div 7 = 10$, remainder 4, so 74 days from today will be 4 days after a Friday, which is a Tuesday.

Problems for Section 5.1 on the Meaning of Division

1. Make up and solve three different story problems (word problems) for $7 \div 3$.

   (a) In the first story problem, the answer should best be expressed as 2, remainder 1.

   (b) In the second story problem, the answer should best be expressed as $2\frac{1}{3}$.

(c) In the third story problem, the answer should best be expressed as 2.33.

2. Explain clearly how to use the *meanings* of multiplication and division as well as the following information in order to determine how many grams 1 cup of water weighs.

1 quart	=	4 cups
1 liter	=	1.056 quarts
1 liter	weighs	1 kilogram
1 kilogram	=	1000 grams

3. Light is known to travel at a speed of about 300,000 kilometers per second. The distance that light travels in one year is called a *light year*. The star Alpha Centauri is 4.34 light years from earth. How many years would it take a rocket travelling at 6000 kilometers per hour to reach Alpha Centauri? (See problem 3 of Chapter 2W.) Solve this problem and explain how you use the meanings of multiplication and division in solving the problem.

4. Susan has a 5 pound bag of flour and an old recipe of her grandmother's calling for one kilogram of flour. She reads on the bag of flour that it weighs 2.26 kilograms. She also reads on the bag of flour that one serving of flour is about $\frac{1}{4}$ cup and that there are about 78 servings in the bag of flour.

   (a) Based on the information above, how many cups of flour should Susan use in her grandmother's recipe? Solve this problem and explain how you use the meanings of multiplication and division in solving the problem.

   (b) How can Susan measure this amount of flour as precisely as possible if she has the following measuring containers available: 1 cup, $\frac{1}{2}$ cup, $\frac{1}{3}$ cup, $\frac{1}{4}$ cup measures, 1 tablespoon? Remember that 1 cup = 16 tablespoons.

5. A certain experiment takes 2 days 5 hours and 14 minutes to perform. A lab must run this experiment 20 times. The lab is set up so that as soon as one experiment is done, the next one will start right away. How long will it take for the 20 experiments to run? Give your answer

in days, hours, minutes. Explain why both multiplication *and* division are needed to solve this problem.

6. Bob wants to estimate 1893 ÷ 275. Bob decides to round 1893 up to 2000. Bob says that since he rounded 1893 *up*, he'll get a better estimate if he rounds 275 in the opposite direction, namely *down* to 250 rather than *up* to 300. Therefore Bob says that 1893 ÷ 275 is closer to 8 (which is 2000 ÷ 250) than to 7, because 2000 ÷ 300 is a little less than 7. This problem will help you investigate whether or not Bob's reasoning is correct.

   (a) Which of 2000 ÷ 250 and 2000 ÷ 300 gives a better estimate to 1893 ÷ 275? Therefore can Bob's reasoning (described above) be correct?

   (b) Use the *meaning* of division (either of the two main interpretations) to explain the following. When estimating the answer to a division problem $A \div B$, if you round A *up*, you will get a better estimate if you also round B *up* rather than *down*. Draw diagrams to aid your explanation.

   (c) Now consider the division problem 1978 ÷ 205. Suppose you round 1978 up to 2000. In this case, will you get a better estimate to 1978 ÷ 205 if you round 205 up to 250 (and compute 2000 ÷ 250) or if you round 205 down to 200 (and compute 2000 ÷ 200)? Reconcile this with your findings in part (b)—you may wish to modify the statement in part (b).

7. You can read on the label of some soda bottles that one liter is 1 quart and 1.8 liquid ounces. Suppose gasoline sells for \$1.35 per gallon. Remember that 1 gallon = 4 quarts, 1 quart = 2 pints, 1 pint = 2 cups, 1 cup = 8 liquid ounces.

   (a) Mentally estimate the price of gas in dollars per liter. Explain your method.

   (b) Find the exact price of gas in dollars per liter. Explain your method.

8. The United States is the second largest country by land area (Russia is first).

(a) Make a guess: about what percent of the surface area of the earth do you think the United States covers?

(b) Now calculate the percent of the earth's surface area that is covered by the United States. The total surface area of the earth is about 5.1×10^8 square kilometers. The United States has an area of about 9.6×10^6 square kilometers.

9. A standard bathtub is approximately $4\frac{1}{2}$ ft long, 2 ft wide, and 1 ft deep. If water comes out of a faucet at the rate of $2\frac{1}{2}$ gallons each minute, how long will it take to fill the bathtub $\frac{3}{4}$ full? You may use the fact that 1 gallon of water occupies .134 cubic feet.

10. In 1994, February 7th and March 7th fell on the same day of the week (namely a Monday). In 1995, February 7th and March 7th fell on the same day of the week (namely a Tuesday). In 1996, February 7th and March 7th did not fall on the same day of the week. (Feb. 7th was a Wednesday and March 7th was a Thursday.) In 1997, February 7th and March 7th fell on the same day of the week (namely, a Friday). Explain, using mathematics and knowledge of our calendar system, why February and March 7th fall on the same day of the week in most years, but don't fall on the same day of the week in some years.

11. October 5, 1998 fell on a Monday. What day of the week did October 5, 1999 fall on? Use division to solve this problem (no peeking at a calendar!). Ok, now you can check your answer on a calendar.

12. Halloween (October 31) of 1998 was on a Saturday, which was great for kids.

(a) How can you use division to determine what day of the week Halloween was on in 1999? (1999 was not a leap year, so Halloween of 1999 was 365 days from Halloween of 1998.)

(b) After 1998, when is the next time that Halloween falls on either a Friday or a Saturday? Again, use mathematics to determine this. Explain your reasoning. (The years 2000, 2004, 2008, 2012 etc. are leap years, so they have 366 days instead of 365.)

13. Must there always be at least one Friday the 13th in every year? Use division to answer this question. (You may answer only for years that

aren't leap years.) To get started on solving this problem, answer the following: If January 13th falls on a Monday, then what day of the week will February 13th, March 13th, etc. fall on? Use division with remainder to answer these questions. Now consider what will happen if January 13th falls on a Tuesday, a Wednesday, etc.

14. Presidents' Day is the third Monday in February. In 2000, Presidents' Day was on February 21. What is the date of Presidents' Day in 2002? Use mathematics to solve this without peeking at a calendar! Explain your reasoning clearly. (Note: the year 2000 was a leap year.)

15. I'm thinking of a number. When you divide it by 2 it has remainder 1, when you divide it by 3 it has remainder 1, when you divide by 4, 5 or 6 it always has remainder 1. The number I am thinking of *could* just be the number 1 (because $1 \div 2 = 0$, remainder 1; $1 \div 3 = 0$, remainder 1; $1 \div 4 = 0$, remainder 1, etc.). Find at least one other such number that is greater than 1.

16. I'm thinking of a number. When you divide it by 12 it has remainder 2 and when you divide it by 16 it also has remainder 2. The number I am thinking of could be the number 2 because $2 \div 12 = 0$, remainder 2 and $2 \div 16 = 0$, remainder 2. Find at least three other such numbers that are greater than 2. How are these numbers related?

17. Three robbers have just acquired a large pile of gold coins. They go to bed, leaving their faithful servant to guard it. In the middle of the night, the first robber gets up, gives two gold coins from the pile to the servant as hush money, divides the remaining pile of gold evenly into three parts, takes one part, and goes back to bed. A little later, the second robber gets up, gives two gold coins from the remaining pile to the servant as hush money, divides the remaining pile evenly into three parts, takes one part, and goes back to bed. A little later, the third robber gets up and does the very same thing. In the morning, when they count up the gold coins, there are 100 of them left. How many were in the pile originally? Explain your answer.

18. (a) Use the meaning of powers of ten to show how to write the expressions in (i), (ii), and (iii) below as a *single* power of ten, i.e., in the form 10^{something}.

i. $10^5 \div 10^2$

ii. $10^6 \div 10^4$

iii. $10^7 \div 10^6$

(b) In each of (i), (ii), and (iii) above, compare the exponents involved. In each case, what is the relationship among the exponents?

(c) Explain why it is always true that $10^A \div 10^B = 10^{A-B}$ when A and B are counting numbers, and A is greater than B.

5.2   Why the Standard Long Division Procedure Works

In Section 4.8 we studied the connection between the *meaning* of multiplication and the standard longhand multiplication *procedure*. We explained why the multiplication procedure gives answers to multiplication problems that agree with the meaning of multiplication. In this section, we will do the same, with division instead of multiplication.

As with multiplication, there is a difference between the *meaning* of division and the longhand *procedure* that is used to calculate answers to division problems. If Tim wants to distribute his prized collection of 102,437 bottle caps equally among his 6 closest friends, how many bottle caps will each friend get? According to the *meaning* of division (with the *how many in each group?* interpretation), each of Tim's friends will get $102,437 \div 6$ bottle caps. Now imagine that Tim carefully divides his bottle cap collection into 6 equal piles, one pile for each friend. Then each pile contains some number of bottle caps. Why is this number of bottle caps the same number that we get when we carry out the standard longhand procedure to divide 102,437 by 6? Why does that long, involved procedure—6 goes into 10 one time, 1 times 6 is 6, ten minus 6 is 4, bring down the 2, etc. — *why does this procedure come up with the actual number of bottle caps that each friend gets?* We will answer this kind of question in this section.

Solving Division Problems Without the Standard Longhand Procedure

In order to analyze why the standard long division procedure works we will first study how to solve division problems *without* the procedure and without a calculator.

In the following example, Vanessa is a student in a combined 3rd and 4th grade class ([50, p.69]):

> *Problem 1: Jesse has 24 shirts. If he puts eight of them in each drawer, how many drawers does he use?*
>
> Vanessa wrote, "$24 - 8 = 16$, $16 - 8 = 8$, $8 - 8 = 0$," and then circled "3" for the answer.
>
> *Problem 2: If Jeremy needs to buy 36 cans of seltzer water for his family and they come in packs of six, how many packs should he buy?*
>
> This time Vanessa added: "$6 + 6 = 12$, $12 + 12 = 24$, $24 + 6 = 30$, $30 + 6 = 36$, ..."

In the first problem, Vanessa starts with 24 and repeatedly *subtracts eights* until she reaches 0. In the second problem, Vanessa repeatedly *adds sixes* until she reaches 36 (the twelves are two sixes and the 24 is 4 sixes because it came from 2 sixes plus 2 sixes). As we'll see later in this section, Vanessa's approach in the first problem is the idea behind the standard division procedure. Vanessa's second approach is equally valid.

Consider another example, where remainders are involved. Suppose that 110 candies will be put into packages of 8 candies each. How many packages can be made, and how many candies will be left over? To solve this, we must calculate $110 \div 8$, which we can do in the following way:

> 10 packages will use up $10 \times 8 = 80$ candies. Then there will be $110 - 80 = 30$ candies left. 3 packages will use $3 \times 8 = 24$ candies. Then there will only be $30 - 24 = 6$ candies left. So $10 + 3 = 13$ packages can be made, and 6 candies will be left over.

Several equations can be used to record this reasoning:

$$
\begin{aligned}
110 - 10 \times 8 &= 30 \\
30 - 3 \times 8 &= 6
\end{aligned}
$$

These can be condensed down to a *single* equation:

$$110 - 10 \times 8 - 3 \times 8 = 6. \tag{5.1}$$

According to the distributive property, this equation can be rewritten as

$$110 - 13 \times 8 = 6. \tag{5.2}$$

Since 6 is less than 8, therefore

$$110 \div 8 = 13, \text{ remainder } 6.$$

The reasoning above used the idea of repeatedly *subtracting* multiples of 8 from 110 in order to calculate $110 \div 8$. Another approach is to repeatedly *add* multiples of 8 until you get as close to 110 as possible without going over. Here's this reasoning with the same problem, $110 \div 8$. It's only slightly different from the reasoning above:

> 10 packages use $10 \times 8 = 80$ candies. 3 packages use $3 \times 8 = 24$ candies. So far, that's $80 + 24 = 104$ candies used. Six more candies makes 110, so you can make 13 packages of candies and 6 will be left over.

As before, we can write a single equation to go along with this reasoning; it involves addition rather than subtraction.

$$10 \times 8 + 3 \times 8 + 6 = 110 \tag{5.3}$$

According to the distributive property, this equation can be rewritten as

$$13 \times 8 + 6 = 110 \tag{5.4}$$

This equation shows that $110 \div 8 = 13$, remainder 6.

Class Activity 5E: Dividing Without Using a Calculator or Long Division

Standard Long Division and the Scaffold Method

As the example on page 337 showed, $110 \div 8$ can be calculated by repeatedly subtracting multiples of 8. The idea of repeated subtraction lies at the heart of the standard long division procedure. In order to understand the workings of the standard long division procedure, we will work with a modification of it, called the *scaffold method*. The scaffold method is less efficient than the standard long division procedure, but its steps are easier to understand.

Below, $4581 \div 7$ is calculated by the standard longhand procedure and by the scaffold method. Both methods arrive at the conclusion that 4581 divided by 7 is 654 with remainder 3.

Standard method :

$$
\begin{array}{r}
654 \\
7 \overline{)4581} \\
-42 \\
\hline
38 \\
-35 \\
\hline
31 \\
-\ 28 \\
\hline
3
\end{array}
$$

Scaffold method :

$$
\begin{array}{rl}
4 & \leftarrow \text{how many 7s are in 31?} \\
50 & \leftarrow \text{how many tens of 7s are in 381?} \\
600 & \leftarrow \text{how many hundreds of 7s are in 4581?} \\
7 \overline{)4581} & \\
-4200 & \\
\hline
381 & \\
-350 & \\
\hline
31 & \\
-\ 28 & \\
\hline
3 &
\end{array}
$$

Notice that the two methods are very similar and accomplish the same thing. The scaffold method works with entire numbers, rather than just portions of numbers (for example, 381 rather than 38). When you use the scaffold method, you have to keep track of place value. For example, in the first step, you ask "how many *hundreds* of sevens are in 4581?", whereas for the standard method you only ask "how many sevens are in 45?". The standard method is really just an abbreviated version of the scaffold method. Therefore, if we can use the *meaning* of division to explain why the scaffold method gives correct answers to division problems, then we will also know why the standard longhand method gives correct answers to division problems.

With the *how many groups?* interpretation of division, $4581 \div 7$ can be thought of as the largest whole number of 7s in 4581, or in other words, the largest whole number that we can multiply 7 by without going over 4581. With that in mind, consider the steps in the scaffold method:

```
         4
        50
       600
  7 )4581
    −4200   ← 600 sevens
      381   ←  what is left over after subtracting 600 sevens
     −350   ← 50 sevens
       31   ← what is left over after subtracting another 50 sevens
      −28   ← 4 sevens
        3   ← what is left over after subtracting another 4 sevens
```

In carrying out the scaffold method, what did we actually do? We started with 4581, subtracted 600 sevens, subtracted another 50 sevens, subtracted another 4 sevens, and in the end, 3 were left over. In other words:

$$4581 - 600 \times 7 - 50 \times 7 - 4 \times 7 = 3. \tag{5.5}$$

Notice that all together, starting with 4581, we subtracted a total of 654 sevens and were left with 3. We come to the same conclusion when we first rewrite the equation

$$4581 - 600 \times 7 - 50 \times 7 - 4 \times 7 = 3$$

as

$$4581 - (600 \times 7 + 50 \times 7 + 4 \times 7) = 3,$$

and then apply the distributive property to get

$$4581 - (600 + 50 + 4) \times 7 = 3,$$

or

$$4581 - 654 \times 7 = 3. \tag{5.6}$$

Equation 5.6 tells us that when we take 654 sevens away from 4581, we are left with 3. Therefore we can conclude that 654 is the largest whole number of 7s in 4581, and therefore $4581 \div 7 = 654$, remainder 3.

In general, why does the scaffold method of division give correct answers to division problems, based on the *meaning* of division? When you solve a division problem $A \div B$ using the scaffold method, you start with the number A, and you repeatedly subtract multiples of B (numbers times B) until a number remains that is less than B. Since you subtracted as many Bs as possible, therefore when you add the total number of Bs that were subtracted, that is the largest whole number of Bs that are in A. What's left over is the remainder. Therefore the answer provided by using the scaffold method to calculate $A \div B$ gives the answer that we expect for $A \div B$ based on the *meaning* of division. Because the standard long division method is just a shorter version of the scaffold method, this also explains why the standard longhand method calculates answers to division problems that agree with the *meaning* of division.

Class Activity 5F: Understanding Long Division

Exercises for Section 5.2 on Why the Standard Long Division Procedure Works

1. The following is one way you could calculate $239 \div 9$.

   > 10 nines is 90. Another 10 nines makes 180. 5 more nines makes 225. 1 more nine makes 234. Five more makes 239. All together that's 26 nines, with 5 left over. So the answer is 26, remainder 5.

   Write a single equation, in the form of Equation 5.1 or Equation 5.3, that incorporates the reasoning above. Use this equation, and the distributive property, to write another equation, like Equation 5.2 or Equation 5.4, that shows the answer to $239 \div 9$.

2. Use the scaffold method to solve the following division problem:

   $31 \overline{)73125}$

3. The following is one way you could calculate $320 \div 17$.

   > 20 seventeens is 340. So 19 seventeens is 17 less, which is 323. Therefore 18 seventeens must be 306. So the answer is 18, remainder 14.

   Write a (nonstandard) scaffold that corresponds to this reasoning. Then write a single equation, similar to Equation 5.1 or Equation 5.3, that incorporates this reasoning. Use this to write another equation, like Equation 5.2 or Equation 5.4, that shows the answer to $320 \div 17$.

4. Use the fact that $13 \times 13 = 169$ and $10 \times 13 = 130$ to calculate $290 \div 13$. Do not use a division procedure and do not use a calculator. Be sure to use both multiplication facts.

5. Calculate $1000 \div 27$ without the use of either a calculator or a division procedure.

6. Suppose you are talking to someone who does not know any division procedure but who knows the meaning of division. Explain to this person why the scaffold method of division gives correct answers to division problems.

Answers to Exercises for Section 5.2 on Why the Standard Long Division Procedure Works

1.

$$10 \times 9 \; + \; 10 \times 9 \; + \; 5 \times 9 \; + \; 1 \times 9 \; + \; 5 \;\; = \;\; 239$$

so, by the distributive property,

$$(10 + 10 + 5 + 1) \times 9 + 5 \;\; = \;\; 239,$$

therefore

$$26 \times 9 + 5 \;\; = \;\; 239,$$

which means

$$239 \div 9 \;\; = \;\; 26, \text{ remainder } 5.$$

2.

$$
\begin{array}{r}
8 \\
50 \\
300 \\
2000 \\
\hline
31\overline{)73125} \\
-62000 \\
\hline
11125 \\
-9300 \\
\hline
1825 \\
-1550 \\
\hline
275 \\
-248 \\
\hline
27
\end{array}
$$

so $73125 \div 31 = 2358$, remainder 27.

3.

$$
\begin{array}{r}
-1 \\
-1 \\
20 \\
\hline
17\overline{)320} \\
-340 \\
\hline
-20 \\
-(-17) \\
\hline
-3 \\
-(-17) \\
\hline
14
\end{array}
$$

$$
\begin{aligned}
20 \times 17 - 1 \times 17 - 1 \times 17 + 14 &= 320 \\
&\text{so} \\
(20 - 1 - 1) \times 17 + 14 &= 320 \\
&\text{therefore} \\
18 \times 17 + 14 &= 320 \\
&\text{which means that} \\
320 \div 17 &= 18, \text{ remainder } 14
\end{aligned}
$$

4. $299 = 169 + 130 = 13 \times 13 + 10 \times 13 = (13 + 10) \times 13 = 23 \times 13$. Therefore since 290 is 9 less than this, 13 goes into 299 twenty two-times. Since 22 thirteens must be thirteen less than 299 (which is 23 thirteens), therefore 290 is four more than 22 thirteens. So $290 \div 13 = 22$, remainder 4.

5. There are many ways you could do this. Here's one. Because $10 \times 27 = 270$, therefore $20 \times 27 = 540$ and $30 \times 27 = 540 + 270 = 810$. Five 27s must be half of 270, which is 135. Therefore

$$
\begin{aligned}
35 \times 27 &= (30 + 5) \times 27 \\
&= 30 \times 27 \ + \ 5 \times 27 \\
&= 810 + 135 \\
&= 945.
\end{aligned}
$$

Two 27s is 54, and adding that to 945 makes 999. Therefore $37 \times 27 = 999$, so $37 \times 27 + 1 = 1000$, and this means that $1000 \div 27 = 37$, remainder 1.

6. See the text.

Problems for Section 5.2 on Why the Standard Long Division Procedure Works

1. Tamarin calculates $834 \div 25$ in the following way:

> I know that four 25s make 100, so I counted 4 for each of the 8 hundreds. This gives me 32. Then there is one more 25 in 34, but there will be 9 left. So the answer is 33, remainder 9.

(a) Explain Tamarin's method in detail and explain why her method is legitimate (do not just state that she gets the correct answer, explain *why* her method gives the correct answer). Include equations as part of your explanation.

(b) Use Tamarin's method to calculate $781 \div 25$.

2. Use some or all of the multiplication facts

$$2 \times 35 = 70,$$
$$10 \times 35 = 350,$$
$$20 \times 35 = 700,$$

repeatedly in order to calculate $2368 \div 35$ without the use of a calculator. Explain your method. Write a non-standard scaffold that corresponds to your method (depending on your method, the arithmetic in your scaffold may or may not be identical to your original arithmetic).

3. Use the two multiplication facts, $30 \times 12 = 360$ and $12 \times 12 = 144$, to calculate $500 \div 12$ without the use of a calculator, standard long division, or the scaffold method of division. Use *both* multiplication facts and explain your method. Give your answer as a whole number with a remainder.

4. (a) Use the scaffold method to calculate $793 \div 4$.

   (b) Interpret each step of the scaffold method in terms of the following story problem: if you have 793 cookies and you want to put them in packages of 4, how many packages will there be, and how many cookies will be left over?

   (c) Write a *single* equation, like Equation 5.5, that incorporates the steps of the scaffold method. Use your equation and the distributive property to write another equation, like Equation 5.6, that shows what the answer to $8321 \div 6$ is. Relate this last equation to portions in the scaffold method.

5. Suppose a student uses the scaffold method in the following non-standard way:

$$
\begin{array}{r}
1 \\
5 \\
10 \\
20 \\
1000 \\
\underline{1000} \\
365\overline{)743425} \\
\underline{365000} \\
378425 \\
\underline{365000} \\
13425 \\
\underline{7300} \\
6125 \\
\underline{3650} \\
2475 \\
\underline{1825} \\
650 \\
\underline{365} \\
285
\end{array}
$$

so $743425 \div 365$ $\quad = 1000 + 1000 + 20 + 10 + 5 + 1$

$\quad = 2036$ remainder 285

Even though this is not standard (even for the scaffold method), is it still mathematically legitimate? Why or why not?

6. A student calculates $6998 \div 7$ as follows, and concludes that $6998 \div 7 = 1000 - 1$, remainder 5, which is 999, remainder 5.

$$
\begin{array}{r}
-1 \\
\underline{1000} \\
7\overline{)6998} \\
\underline{7000} \\
-2 \\
\underline{-7} \\
5
\end{array}
$$

While this is obviously not the standard way to use the scaffold method, explain why the method does correspond to legitimate reasoning.

5.3   Fractions and Division

In chapter 2, we simply stated that a fraction is equal to its numerator divided by its denominator:

$$\frac{A}{B} = A \div B.$$

In fact, instead of writing the "divided by" symbol, $\div$, we sometimes write /, so that $2 \div 5$ can also be expressed as 2/5. But the expression 2/5 is also another way to write $\frac{2}{5}$. So the notation we use equates fractions with division. In this section we will discuss why it is legitimate to equate fractions with division. By expressing fractions in terms of division, we can make sense of fractions that have negative numerators and/or denominators.

Class Activity 5G: Can Fractions be Defined in a Different Way?

Explaining Why Fractions Can be Expressed in Terms of Division

According to our definitions of fractions and of division, the expressions

$$\frac{2}{5} \quad \text{and} \quad 2 \div 5$$

do have different meanings:

$\frac{2}{5}$ of a pie is the amount of pie formed by 2 pieces when the pie is divided into 5 equal pieces.

$2 \div 5$ is the amount of pie each person will receive if 2 (identical) pies are divided equally among 5 people (using the *how many in each group?* interpretation).

Notice the difference: $\frac{2}{5}$ refers to *2 pieces of pie*, whereas $2 \div 5$ refers to dividing *2 pies*. But is it the same amount of pie either way? To divide 2 pies equally among 5 people you can divide each pie into 5 equal pieces and give each person one piece from each pie, as shown in Figure 5.5.

One person's share of pie consists of 2 pieces of pie, and each of those pieces is $\frac{1}{5}$ of a pie, i.e., each piece comes from a pie that has been divided into 5 equal pieces. So each of the 5 people sharing the 2 pies gets $\frac{2}{5}$ of a pie.

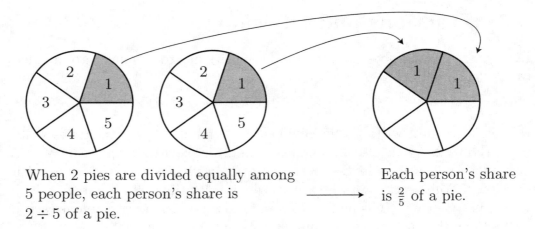

When 2 pies are divided equally among 5 people, each person's share is $2 \div 5$ of a pie. $\longrightarrow$ Each person's share is $\frac{2}{5}$ of a pie.

Figure 5.5: Dividing 2 Pies Equally Among 5 People

So one person's share of pie can be described in two ways: as $2 \div 5$ of a pie and as $\frac{2}{5}$ of a pie. Therefore these two ways of describing a person's share of pie must be equal:

$$2 \div 5 = \frac{2}{5}.$$

The discussion above applies equally well when other whole numbers replace 2 and 5 (except that 5 should not be replaced with 0). So in general, if A and B are whole numbers and B is not 0, then $A \div B$ really is equal to $\frac{A}{B}$.

Decimal Representations of Fractions

Every fraction $\frac{A}{B}$ of integers can be plotted on a number line and therefore can be expressed as a decimal. To express a fraction as a decimal, we use the fact that

$$\frac{A}{B} = A \div B,$$

and we divide the numerator by the denominator. For example,

$$\begin{array}{rcl} \frac{5}{16} & = & 5 \div 16 = .3125, \\ \frac{1}{12} & = & 1 \div 12 = .083333333\ldots, \\ \frac{2}{7} & = & 2 \div 7 = .285714285714\ldots. \end{array}$$

Using Long Division to Express a Fraction as a Decimal

When we divide the numerator of a fraction by the denominator of the fraction longhand in order to write the fraction as a decimal, we use an extension of the long division procedure that we studied in Section 5.2. By working with the scaffold method, we can see why it makes sense to use this procedure to express fractions as decimals.

For example, how do we write

$$\frac{243}{7}$$

as a decimal? First, start to calculate

$$243 \div 7$$

as before:

```
        4  ← how many 7s are in 33?
       30  ← how many tens of 7s are in 243?
   7)243
      210
       33
       28
        5
```

Instead of stopping at this point and saying that $243 \div 7 = 34$, remainder 5, we must to continue, and we must divide 5 by 7. In order to continue, we must ask about tenths, hundredths, and thousandths of 7s, not just tens of

7s and ones of 7s:

$$
\begin{array}{r}
.004 \\
.01 \\
.7 \\
4 \\
\underline{30} \\
7)\overline{243} \\
\underline{210} \\
33 \\
\underline{28} \\
5 \\
\underline{4.9} \\
.1 \\
\underline{.07} \\
.03 \\
\underline{.028} \\
.002
\end{array}
\quad
\begin{array}{l}
\leftarrow \text{ how many thousandths of 7s are in .03?} \\
\leftarrow \text{ how many hundredths of 7s are in .1?} \\
\leftarrow \text{ how many tenths of 7s are in 5?} \\
\leftarrow \text{ how many 7s are in 33?} \\
\leftarrow \text{ how many tens of 7s are in 243?}
\end{array}
$$

Therefore

$$\frac{243}{7} = 243 \div 7 = 34.714\ldots$$

If we wanted to show more decimal places in the decimal representation of $\frac{243}{7}$ we would simply continue the long division process.

Using Diagrams to Show Decimal Representations of Fractions

A good way to get a feel for the relationship between fractions and decimals is to use diagrams to determine the decimal representations of fractions.

For example, Figure 5.6 shows that

$$\frac{1}{4} = .25$$

in the following way. Consider the large square (made up of 100 small squares) as representing 1. Then each of 4 equal pieces of this large square represents $\frac{1}{4} = 1 \div 4$. Therefore $\frac{1}{4}$ is represented by the shaded portion of Figure 5.6, which consists of vertical strips of 10 small squares, and 5 more small squares. Becuse each vertical strip of 10 small squares represents $\frac{1}{10}$

and each small square represents $\frac{1}{100}$, therefore Figure 5.6 shows that

$$\frac{1}{4} = 2 \cdot \frac{1}{10} + 5 \cdot \frac{1}{100} = .25.$$

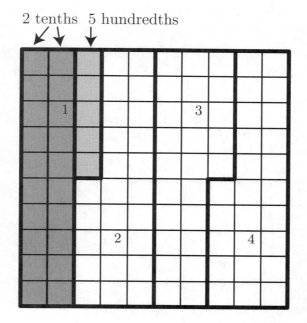

Figure 5.6: $\frac{1}{4} = 2 \cdot \frac{1}{10} + 5 \cdot \frac{1}{100} = .25$

The case of $\frac{1}{3}$ is a bit more subtle. Once again, consider a large square made up of 100 small squares, and let the large square represent 1. To show the decimal representation of $\frac{1}{3}$ we must express $\frac{1}{3}$ of the large square in terms of strips of 10 small squares, and individual small squares. Since each vertical strip of 10 small squares represents $\frac{1}{10}$, and since each individual small square represents $\frac{1}{100}$, we will then have expressed $\frac{1}{3}$ in terms of tenths and hundredths, which will tell us the decimal representation of $\frac{1}{3}$ to 2 decimal places.

To divide the large square into 3 equal pieces, start by making 3 groups of 3 vertical strips of 10 squares, as in Figure 5.7. This almost divides the whole square into 3 equal pieces, except that there is a strip of 10 small squares left that must also be divided into 3 equal pieces in order to divide the whole square into 3 equal pieces. We can make 3 groups of 3 squares and that almost divides the strip of 10 small squares into 3 equal pieces, except that there is still one small square remaining that also needs to be divided into 3 equal pieces. What does this tell us? Collecting up pieces, we see that

$$\frac{1}{3} = 3 \cdot \frac{1}{10} + 3 \cdot \frac{1}{100} + \text{ a little more}$$

where the "little more" is less than $\frac{1}{100}$ (in fact it is exactly $\frac{1}{3}$ of $\frac{1}{100}$). Therefore

$$\frac{1}{3} = .33\ldots.$$

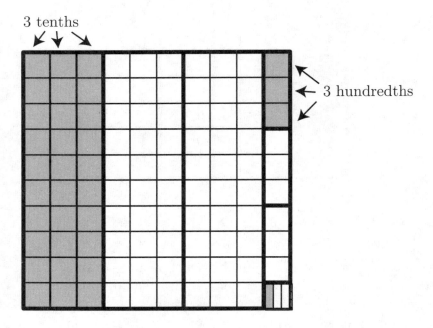

Figure 5.7: Showing The Decimal Representation Of $\frac{1}{3}$

To determine more digits in the decimal representation of $\frac{1}{3}$ we'd need to subdivde the small remaining square into 3 equal parts, and this would lead to doing the original procedure above with the large square over again, as indicated by Figure 5.8. Therefore the digits in the thousandths and ten-thousandths places are also 3s. If we imagine continuing this process forever, we see that all the digits in the decimal representation of $\frac{1}{3}$ are 3s:

$$\frac{1}{3} = .33333\ldots,$$

the 3s go on forever.

Class Activity 5H: Showing Decimal Representations of Fractions

Fractions With Negative Numerators or Denominators

How can we make sense of fractions such as $\frac{-13}{5}$ or $\frac{13}{-5}$ or $\frac{-13}{-5}$? Above, we explained why it makes sense that when A and B are counting numbers,

$$\frac{A}{B} = A \div B.$$

We can use this relationship to *define* fractions that have negative numerators and/or denominators. Therefore, if A and B are whole numbers and B is not zero, then:

$$\begin{array}{ccccccc}
\frac{-A}{B} & = & (-A) \div B & = & -(A \div B) & = & -\frac{A}{B}, \\
\frac{A}{-B} & = & A \div (-B) & = & -(A \div B) & = & -\frac{A}{B}, \\
\frac{-A}{-B} & = & (-A) \div (-B) & = & A \div B & = & \frac{A}{B}.
\end{array}$$

So

$$\frac{-13}{5} = -\frac{13}{5}$$
$$\frac{13}{-5} = -\frac{13}{5}$$
$$\frac{-13}{-5} = \frac{13}{5}$$

Figure 5.9 shows $\frac{-13}{5}$ plotted on a number line. According to the rules of Section 2.3 for locating numbers on number lines, because

$$\frac{-13}{5} = -\frac{13}{5},$$

therefore $\frac{-13}{5}$ should be located to the left of 0, at a distance of $\frac{13}{5}$ units away from 0.

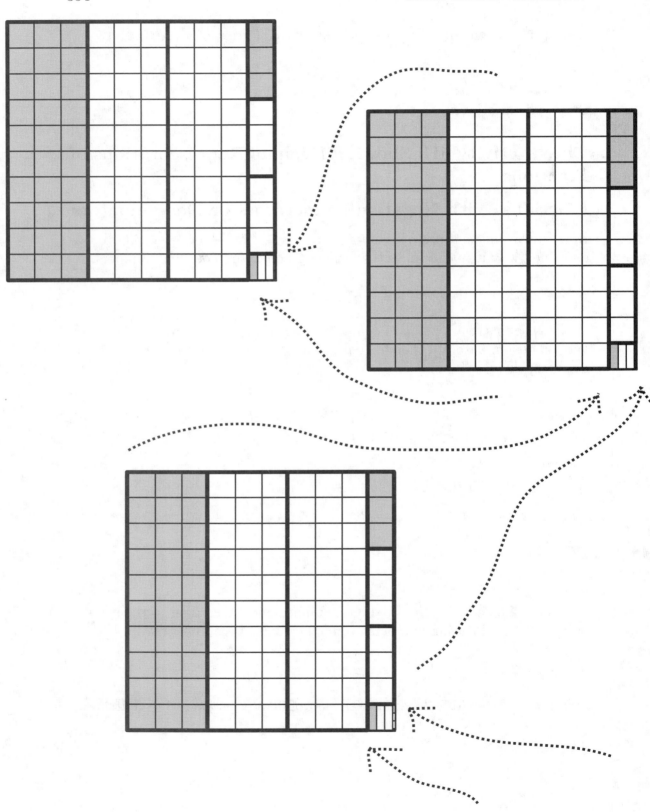

Figure 5.8: $\frac{1}{3} = .333333\ldots$

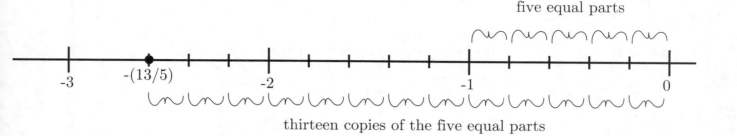

five equal parts

thirteen copies of the five equal parts

Figure 5.9: $\frac{-13}{5} = -\frac{13}{5}$ On A Number Line

Exercises for Section 5.3 on Fractions and Division

1. Use the meaning of fractions and the meaning of division to explain why $\frac{3}{10} = 3 \div 10$. Your explanation should be "general", in the sense that you could see why $\frac{3}{10} = 3 \div 10$ would still be true if other numbers were to replace 3 and 10.

2. Use the large square in Figure 5.10 (which is subdivided into 100 small squares) to help you explain why the decimal representation of $\frac{1}{8}$ is .125.

3. Plot $\frac{-11}{6}$ on a number line and explain why it should be plotted there.

Answers to Exercises for Section 5.3 on Fractions and Division

1. If there are 3 (identical) pies to be divided equally among 10 people, then according to the meaning of division (with the *how many in each group?* interpretation) each person will get $3 \div 10$ of a pie. To divide the pies, you can divide each pie into 10 pieces and give each person 1 piece from each of the 3 pies. One person's share is shown shaded in Figure 5.11. Each person then gets 3 pieces, where each piece is $\frac{1}{10}$ of a pie, i.e., each piece is 1 part when the pie is divided into 10 equal parts. Therefore each person gets $\frac{3}{10}$ of a pie according to the meaning of fractions. Because each person's share can be described both as $3 \div 10$ of a pie and as $\frac{3}{10}$ of a pie, therefore $\frac{3}{10} = 3 \div 10$.

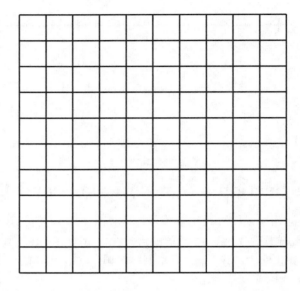

Figure 5.10: Show $\frac{1}{8}$

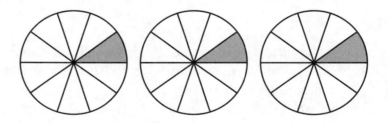

Figure 5.11: One Person's Share When 3 Pies are Divided Among 10 People

2. By thinking of the large square as representing 1, each small square then represents $\frac{1}{100}$. To divide the large square into 8 equal pieces, first make 8 strips of 10 small squares. Then there are still 20 small squares left to be divided into 8 equal pieces. You can make 8 groups of 2 small squares from these 20 small squares. That leaves 4 small squares left to be divided into 8 equal pieces. Dividing each small square in half makes 8 half squares. All together, $\frac{1}{8}$ of the big square is then obtained by collecting up these various parts: one strip of 10 small squares, 2 small squares, and one half of a small square. So Figure 5.12 shows that

$$\frac{1}{8} = 1 \cdot \frac{1}{10} + 2 \cdot \frac{1}{100} + 5 \cdot \frac{1}{1000}$$
$$= .125$$

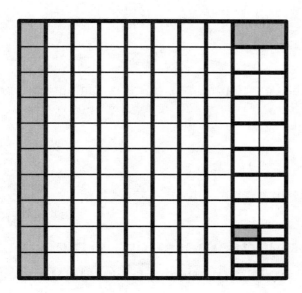

Figure 5.12: Showing $\frac{1}{8} = .125$

3. Because

$$\frac{-11}{6} = -11 \div 6 = -(11 \div 6) = -\frac{11}{6},$$

therefore we should plot $\frac{-11}{6}$ to the left of 0, at a distance of $\frac{11}{6}$ units from 0, as shown in Figure 5.13.

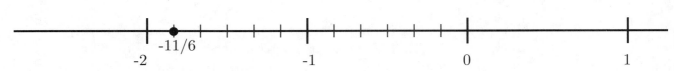

Figure 5.13: Plotting $\frac{-11}{6} = -\frac{11}{6}$

Problems for Section 5.3 on Fractions and Division

1. The text presented one way to explain why $\frac{A}{B} = A \div B$. This problem will help you explain in a different way why fractions can be expressed in terms of division.

   (a) Use the meaning of multiplication, the meaning of fractions, and the meaning of division to explain why

   $$\frac{1}{10} \cdot 3 = 3 \div 10.$$

   (b) Use the equation in part (a), the commutative property of multiplication, and the meaning of fractions to explain why

   $$3 \div 10 = \frac{3}{10}.$$

2. Use a diagram to show why $\frac{1}{9} = .11\ldots.$.

5.4   Relating Remainder, Mixed Number, and Decimal Answers to Whole Number Division Problems

As was discussed at the beginning of this chapter, the answer to a whole number division problem can be given in one of the following forms:

a whole number with a remainder,

a mixed number,

a decimal.

In this section we will discuss how the three different types of answers to a whole number division problem are related.

Consider the division problem

$$29 \div 6.$$

We can answer this division problem in any of the following three ways:

$$4, \text{ remainder } 5, \quad 4\frac{5}{6}, \quad \text{or } 4.833\ldots.$$

How are these related? The last two answers are different forms of the same number: the decimal representation of $\frac{5}{6}$ is $.833\ldots$, which we can determine by calculating $5 \div 6$ as discussed in Section 5.3. Therefore

$$4\frac{5}{6} = 4.833\ldots.$$

The same will be true for other division problems with whole numbers: the mixed number answer and the decimal answer are equal to each other, they are just different forms of the same number.

How are the remainder answer and the mixed number answer related? For the division problem $29 \div 6$, the remainder is 5, the divisor is 6, and the fractional part of the mixed number answer, $4\frac{5}{6}$, is $\frac{5}{6}$, which is in the form

$$\frac{\text{remainder}}{\text{divisor}}.$$

This will be the case in general. Why does this make sense? The fractional part of the mixed number answer comes from dividing the remainder, as we will see in the following example. Suppose we have 29 pizzas to be divided equally among 6 classrooms. How many pizzas does each classroom get? Since $29 \div 6 = 4$, remainder 5, we can say that each classroom will get 4 pizzas, and there will be 5 pizzas left over. But instead of just giving away the remaining 5 pizzas, we might want to divide these 5 pizzas into 6 equal parts. When 5 pizzas are divided equally among 6 classrooms, how much pizza does each classroom get? Each classroom gets

$$5 \div 6 = \frac{5}{6}$$

of a pizza, according to Section 5.3, as we also see in Figure 5.14.

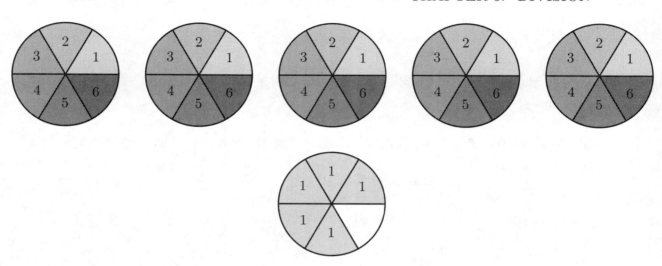

Figure 5.14: Dividing 5 Pizzas Among 6 Classrooms

So instead of giving each class 4 pizzas and having 5 pizzas left over, we can give each class $4\frac{5}{6}$ of a pizza. In other words,

$$29 \div 6 = 4, \text{ remainder } 5$$

is equivalent to

$$29 \div 6 = 4\frac{5}{6}.$$

Notice that in the mixed number version, the remainder 5 becomes the numerator of the fractional part of the answer, while the denominator is the divisor, 6.

The same reasoning works with other numbers. If A and B are natural numbers, then

$$A \div B = Q, \text{ remainder } R$$

is equivalent to

$$A \div B = Q\frac{R}{B}.$$

In the mixed number answer to $A \div B$, the fractional part of of the answer is the remainder R over the divisor, B.

Class Activity 5I: Relating Remainder, Mixed Number, and Decimal Answers to Division Problems

Exercises for Section 5.4 on Relating Different Answers to Division Problems

1. Explain how the remainder and mixed number answers to $14 \div 3$ are related, and explain why this is so.

Answers to Exercises for Section 5.4

1. The "remainder" answer to $14 \div 3$ is 4, remainder 2. The mixed number answer is $4\frac{2}{3}$. In general, to get the fractional part of the mixed number answer you make a fraction whose numerator is the remainder and whose denominator is the divisor. To see why this makes sense, consider an example. If you have 14 cookies to be divided equally among 3 people, then you could give each person 4 cookies and have 2 cookies left over, or you could take those 2 remaining cookies and divide them equally among the 3 people. You could do this by dividing each cookie into 3 equal parts and giving each person 2 parts (as you could show in a picture like Figure 5.14). In this way, each person would get $4\frac{2}{3}$ cookies.

Problems for Section 5.4

1. Jessica calculates that $7 \div 3 = 2$, remainder 1. When Jessica is asked to write her answer as a decimal, she simply puts the remainder 1 behind the decimal point:
$$7 \div 3 = 2.1$$
Is Jessica correct or not? If not, explain clearly why not and help Jessica understand the correct answer. Draw diagrams to aid your explanation.

2. When you divide whole numbers using an ordinary calculator, the answer is displayed as a decimal. But what if you want the answer as a whole number with a remainder? Here's a method to determine the remainder with an ordinary calculator, illustrated with the example of $236 \div 7$.

Use the calculator to divide the whole numbers.

$$236 \div 7 = 33.71428\ldots.$$

Subtract off the whole number part of the answer.

$$33.71428\ldots - 33 = .71428\ldots.$$

Multiply the resulting decimal by the number you divided by (which was 7 here).

$$.71428\ldots \times 7 = 4.99999.$$

Round the resulting number to the nearest whole number. This is your remainder. In this example the remainder is 5. So $236 \div 7 = 33$, remainder 5.

(a) Solve at least two more division problems with whole numbers and use the above method to find the answer as a whole number with a remainder. Check that your answers are correct.

(b) Now solve the same division problems you did in part (a), except this time, give your answers as mixed numbers instead of as decimals.

(c) Use the mixed number version to help you explain why the calculator method (above) for determining the remainder works.

5.5   Dividing Fractions

In this section, we will discuss the two interpretations of division for fractions, and we will see why the standard "invert and multiply" *procedure* for dividing fractions gives answers to fraction division problems that agree with what we expect from the *meaning* of division.

Let's review the meaning of division, now in the context of fractions.

The *how many groups?* interpretation:

With the *how many groups?* interpretation of division,

$$\frac{5}{2} \div \frac{2}{3}$$

means the number of groups when $\frac{5}{2}$ of an object is divided into groups with $\frac{2}{3}$ of an object in each group. For example, suppose you are making popcorn balls and each popcorn ball requires $\frac{2}{3}$ of a cup of popcorn. If you have $2\frac{1}{2} = \frac{5}{2}$ of a cup of popcorn, then how many popcorn balls can you make? In this case you want to divide $\frac{5}{2}$ of a cup of popcorn into groups (balls) with $\frac{2}{3}$ of a cup of popcorn in each group. According to the *how many groups?* interpretation of division, you can make

$$\frac{5}{2} \div \frac{2}{3}$$

popcorn balls.

The *how many in each group?* interpretation:

With the *how many in each group?* interpretation of division,

$$\frac{3}{4} \div \frac{1}{2}$$

is the number of objects in each group when $\frac{3}{4}$ of an object is distributed equally among $\frac{1}{2}$ of a group. A clearer way to say this is: $\frac{3}{4} \div \frac{1}{2}$ is the number of objects (or fraction of an object) in one group when $\frac{3}{4}$ of an object is put into $\frac{1}{2}$ of a group. For example, suppose you pour $\frac{3}{4}$ of a pint of blueberries into a container and this fills $\frac{1}{2}$ of the container. How many pints of blueberries will it take to fill the whole container? In this case, $\frac{3}{4}$ of a pint of blueberries is put into (i.e., distributed equally among) $\frac{1}{2}$ of a group (container). So according to the *how many in each group?* interpretation of division, the number of pints of blueberries in one group (one container) is

$$\frac{3}{4} \div \frac{1}{2}.$$

Although most people prefer the *how many in each group?* interpretation of division when whole numbers are involved, many people find this interpretation of division difficult to grasp when fractions are involved. It might help you to understand it better if you think about replacing the fractions with whole numbers. For example, if you have 3 pints of blueberries and this fills up 2 containers, then how many pints of blueberries are in each container? This is solved by $3 \div 2$ according to the *how many in each group?* interpretation. Therefore if you replace the 3 pints with $\frac{3}{4}$ of a pint, and the

2 containers with $\frac{1}{2}$ of a container, the problem should be solved in the same way as before: $3 \div 2$ now becomes $\frac{3}{4} \div \frac{1}{2}$.

Here is another way to think about the same problem. Because $\frac{1}{2}$ of the container is filled, and because this amount is $\frac{3}{4}$ of a pint, therefore $\frac{1}{2}$ of the number of pints in a full container is $\frac{3}{4}$ of a pint. In other words:

$$\frac{1}{2} \times \text{number of pints in full container} = \frac{3}{4}.$$

Therefore

$$\text{number of pints in full container} = \frac{3}{4} \div \frac{1}{2}.$$

Dividing *By* $\frac{1}{2}$ Versus Dividing *In* $\frac{1}{2}$

In mathematics, language is used much more precisely and carefully than in everyday conversation. This is one source of difficulty in learning mathematics. For example, consider the two phrases:

> dividing *by* $\frac{1}{2}$,
>
> dividing *in* $\frac{1}{2}$.

You may feel that these two phrases mean the same thing, however, mathematically, they do not. To divide a number, say 5, *by* $\frac{1}{2}$ means to calculate $5 \div \frac{1}{2}$. Remember that we read $A \div B$ as A divided *by* B. We would divide 5 *by* $\frac{1}{2}$ if we wanted to know how many half cups of flour are in 5 cups of flour, for example. (Notice that there are 10 half-cups of flour in 10 cups of flour, not $2\frac{1}{2}$.)

On the other hand, to divide a number *in* half means to find *half of that number*. So to divide 5 *in* half means to find $\frac{1}{2}$ *of* 5. One half *of* 5 means $\frac{1}{2} \times 5$. So dividing *in* $\frac{1}{2}$ is the same as dividing *by* 2.

Class Activity 5J: Are These Division Problems?

The "Invert and Multiply" Procedure for Fraction Division

Although division with fractions can be difficult to interpret, the *procedure* for dividing fractions is quite easy. To divide fractions, such as

$$\frac{3}{4} \div \frac{2}{3} \quad \text{and} \quad 6 \div \frac{2}{5},$$

we can use the familiar "invert and multiply" method in which we invert the divisor and multiply by it:

$$\frac{3}{4} \div \frac{2}{3} = \frac{3}{4} \times \frac{3}{2} = \frac{3 \cdot 3}{4 \cdot 2} = \frac{9}{8},$$

and

$$6 \div \frac{2}{5} = \frac{6}{1} \div \frac{2}{5} = \frac{6}{1} \times \frac{5}{2} = \frac{6 \cdot 5}{1 \cdot 2} = \frac{30}{2} = 15.$$

Another way to describe this "invert and multiply" method for dividing fractions is in terms of the *reciprocal* of the divisor. The **reciprocal** of a fraction $\frac{A}{B}$ is the fraction $\frac{B}{A}$. In order to divide fractions, we should *multiply by the reciprocal of the divisor.*

The procedure for dividing fractions is easy enough to carry out, but why is it a valid method? Before we consider this question in general, consider a special case. Recall that every whole number is equal to a fraction (for example, $6 = \frac{6}{1}$). Therefore the "invert and multiply" procedure for dividing fractions can be applied to whole numbers as well as to fractions. According to the "invert and multiply" procedure,

$$2 \div 3 = \frac{2}{1} \div \frac{3}{1} = \frac{2}{1} \cdot \frac{1}{3} = \frac{2 \cdot 1}{1 \cdot 3} = \frac{2}{3}.$$

Notice that this result, that $2 \div 3 = \frac{2}{3}$, agrees with our findings earlier in this chapter: that fractions can be described in terms of division, namely that $\frac{A}{B} = A \div B$.

In general, why is the "invert and multiply" procedure a valid way to divide fractions? One way to explain this is to relate fraction division to fraction multiplication. Recall that every division problem is equivalent to a multiplication problem (actually two multiplication problems):

$$A \div B = ?$$

is equivalent to

$$? \cdot B = A$$

(or $B \cdot ? = A$). So

$$\frac{3}{4} \div \frac{2}{3} = ?$$

is equivalent to

$$? \cdot \frac{2}{3} = \frac{3}{4}. \tag{5.7}$$

Now remember that we want to see if the "invert and multiply" rule for fraction division makes sense. This rule says that $\frac{3}{4} \div \frac{2}{3}$ ought to be equal to

$$\frac{3 \cdot 3}{4 \cdot 2}.$$

Let's check that this fraction works in the place of the ? in Equation 5.7. In other words, let's check that if we multiply $\frac{3 \cdot 3}{4 \cdot 2}$ times $\frac{2}{3}$, then we really do get $\frac{3}{4}$:

$$\frac{3 \cdot 3}{4 \cdot 2} \cdot \frac{2}{3} = \frac{3 \cdot 3 \cdot 2}{4 \cdot 2 \cdot 3} = \frac{3 \cdot (3 \cdot 2)}{4 \cdot (2 \cdot 3)} = \frac{3 \cdot (3 \cdot 2)}{4 \cdot (3 \cdot 2)} = \frac{3}{4}.$$

Therefore the answer given by the "invert and multiply" procedure really is the answer to the original division problem $\frac{3}{4} \div \frac{2}{3}$. Notice that the line of reasoning above applies in the same way when other fractions replace the fractions $\frac{2}{3}$ and $\frac{3}{4}$ used above.

It will still be valuable to explore fraction division further, interpreting fraction division directly rather than through multiplication.

Class Activity 5K: Understanding Fraction Division

Why The *Invert And Multiply* Procedure for Dividing Fractions is Valid

Above, we explained why the "invert and multiply" procedure for dividing fractions is valid by considering fraction division in terms of fraction multiplication. Now we will explain why the "invert and multiply" procedure is valid by working directly with the meaning of division.

Consider the division problem

$$\frac{2}{3} \div \frac{1}{2}.$$

The following is a story problem for this division problem:

How many $\frac{1}{2}$ cups of water are in $\frac{2}{3}$ of a cup of water?

Or, said another way:

How many times will we need to pour $\frac{1}{2}$ cup of water into a container that holds $\frac{2}{3}$ cup of water in order to fill the container?

From the diagram in Figure 5.15 we can say right away that the answer to this problem is "one and a little more" because one half cup clearly fits in two thirds of a cup, but then a little more is still needed to fill the two thirds of a cup. But what is this "little more?" Remember the original question: we want to know how many $\frac{1}{2}$ cups of water are in $\frac{2}{3}$ of a cup of water. So the answer should be of the form "so and so many $\frac{1}{2}$ cups of water." This means that we need to express this "little more" as *a fraction of* $\frac{1}{2}$ *cup of water.* How can we do that? By subdividing both the $\frac{1}{2}$ and the $\frac{2}{3}$ into common parts, namely by using common denominators.

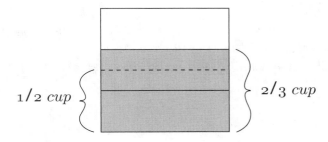

Figure 5.15: How many 1/2 cups of water are in 2/3 cup?

When we give $\frac{1}{2}$ and $\frac{2}{3}$ the common denominator of 6, then, as in Figure 5.16, the $\frac{1}{2}$ cup of water is made out of 3 parts (3 sixths of a cup of water), and the $\frac{2}{3}$ cup of water is made out of 4 parts (4 sixths of a cup of water), so the "little more" we were discussing above is just one of those parts. Since $\frac{1}{2}$ cup is 3 parts, and the "little more" is 1 part, the "little more" is $\frac{1}{3}$ *of the* $\frac{1}{2}$ *cup of water.* This explains why $\frac{2}{3} \div \frac{1}{2} = 1\frac{1}{3}$: there's a whole $\frac{1}{2}$ cup plus another $\frac{1}{3}$ of the $\frac{1}{2}$ cup in $\frac{2}{3}$ of a cup of water.

To recap: we are considering the fraction division problem $\frac{2}{3} \div \frac{1}{2}$ in terms of the story problem "how many $\frac{1}{2}$ cups of water are in $\frac{2}{3}$ of a cup of water?" If we give $\frac{1}{2}$ and $\frac{2}{3}$ the common denominator of 6, then we can rephrase the problem as "how many $\frac{3}{6}$ of a cup are in $\frac{4}{6}$ of a cup?" But in terms of Figure 5.16, this is equivalent to the problem "how many 3s are in 4?" which is the problem $4 \div 3$, whose answer is $\frac{4}{3}$. Notice that $\frac{4}{3}$ is exactly the same answer given by the "invert and multiply" procedure for fraction division:

$$\frac{2}{3} \div \frac{1}{2} = \frac{2}{3} \cdot \frac{2}{1} = \frac{2 \cdot 2}{3 \cdot 1} = \frac{4}{3}.$$

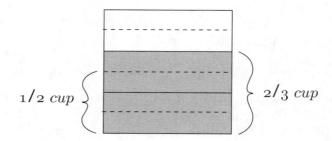

Figure 5.16: How many 3/6 cups of water are in 4/6 cups?

So the "invert and multiply" procedure gives the same answer to $\frac{2}{3} \div \frac{1}{2}$ that we arrive at by using the meaning of division.

The same line of reasoning will work for any fraction division problem

$$\frac{A}{B} \div \frac{C}{D}.$$

Thinking logically, as above, and interpreting $\frac{A}{B} \div \frac{C}{D}$ as "how many $\frac{C}{D}$ cups of water are in $\frac{A}{B}$ cups of water?", one can conclude that

$$\frac{A}{B} \div \frac{C}{D} = \frac{A \cdot D}{B \cdot D} \div \frac{B \cdot C}{B \cdot D} = \frac{A \cdot D}{B \cdot C}.$$

The final expression, $\frac{A \cdot D}{B \cdot C}$, is the answer provided by the "invert and multiply" procedure for dividing fractions. Therefore we know that the "invert and multiply" procedure gives answers to division problems that agree with what we expect from the meaning of division.

Exercises for Section 5.5 on Dividing Fractions

1. Which of the following are solved by the division problem $\frac{3}{4} \div \frac{1}{2}$? For those that are, which interpretation of division is used? For those that are not, determine how to solve the problem, if it can be solved.

   (a) $\frac{3}{4}$ of a bag of gummi worms make $\frac{1}{2}$ a cup. How many cups of gummi worms are in one bag?

   (b) $\frac{3}{4}$ of a bag of gummi worms make $\frac{1}{2}$ a cup. How many bags of gummi worms does it take to make one cup?

(c) You have $\frac{3}{4}$ of a bag of gummi worms and a recipe that calls for $\frac{1}{2}$ of a cup of gummi worms. How many batches of your recipe can you make?

(d) You have $\frac{3}{4}$ of a cup of gummi worms and a recipe that calls for $\frac{1}{2}$ of a cup of gummi worms. How many batches of your recipe can you make?

(e) If $\frac{3}{4}$ of a pound of candy costs $\frac{1}{2}$ of a dollar, then how many pounds of candy should you be able to buy for 1 dollar?

(f) If you have $\frac{3}{4}$ of a pound of candy and you divide the candy in $\frac{1}{2}$, then how much candy will you have in each portion?

(g) If $\frac{1}{2}$ of a pound of candy costs \$1, then how many dollars should you expect to pay for $\frac{3}{4}$ of a pound of candy?

2. Frank, John, and David earned \$14 together. They want to divide it equally, except that David should only get a half share, since he did half as much work as either Frank or John did (and Frank and John worked equal amounts). Write a division problem to figure out how much Frank should get. Which interpretation of division does this use?

3. Bill leaves a tip of \$4.50 for a meal. If the tip is 15% of the cost of the meal, then how much did the meal cost? Write a division problem to solve this. Which interpretation of division does this use?

4. Compare the arithmetic needed to solve the following problems.

(a) What fraction of a $\frac{1}{3}$ cup measure is filled when we pour in $\frac{1}{4}$ cup of water?

(b) What is one quarter of $\frac{1}{3}$ cup?

(c) How much more is $\frac{1}{3}$ cup than $\frac{1}{4}$ cup?

(d) If $\frac{1}{4}$ cup of water fills $\frac{1}{3}$ of a plastic container, then how much water will the full container hold?

5. Write a simple story problem for $1 \div \frac{5}{7}$. Use the story problem and a diagram to help you solve the problem.

6. Use the meanings of multiplication and division to solve the following problems.

(a) Suppose you drive 4500 miles every half year in your car. At the end of 3 and $\frac{3}{4}$ years, how many miles will you have driven?

(b) Mo used 128 ounces of liquid laundry detergent in 6 and $\frac{1}{2}$ weeks. If Mo continues to use laundry detergent at this rate, how much will he use in a year?

(c) Suppose you have a 32 ounce bottle of weed killer concentrate. The directions say to mix two and a half ounces of weed killer concentrate with enough water to make a gallon. How many gallons of weed killer will you be able to make from this bottle?

7. The line segment below is $\frac{2}{3}$ of a unit long. Show a line segment that is $\frac{5}{2}$ of a unit long. Explain how this problem is related to fraction division.

$$\overline{}$$
$\frac{2}{3}$ unit

Answers To Exercises For Section 5.5 on Dividing Fractions

1. (a) This can be rephrased as "if $\frac{1}{2}$ of a cup of gummi worms are put into $\frac{3}{4}$ of a bag, then how many cups are in one bag?", therefore this is a *how many in each group?* division problem illustrating $\frac{1}{2} \div \frac{3}{4}$, not $\frac{3}{4} \div \frac{1}{2}$.

(b) This is solved by $\frac{3}{4} \div \frac{1}{2}$, according to the *how many in each group?* interpretation. A group is a cup and each object is a bag of gummi worms.

(c) This can't be solved because you don't know how many cups of gummi worms are in $\frac{3}{4}$ of a bag.

(d) This is solved by $\frac{3}{4} \div \frac{1}{2}$, according to the *how many groups?* interpretation. Each group consists of $\frac{1}{2}$ of a cup of gummi worms.

(e) This is solved by $\frac{3}{4} \div \frac{1}{2}$, according to the *how many in each group?* interpretation. This is because you can think of the problem as saying that $\frac{3}{4}$ of a pound of candy is put into $\frac{1}{2}$ of a group and you want to know how many pounds are in 1 group.

(f) This is solved by $\frac{3}{4} \times \frac{1}{2}$, not $\frac{3}{4} \div \frac{1}{2}$. It is dividing *in half*, not dividing *by* $\frac{1}{2}$.

(g) This is solved by $\frac{3}{4} \times \frac{1}{2}$, according to the *how many groups?* interpretation because you want to know how many $\frac{1}{2}$ pounds are in $\frac{3}{4}$ of a pound. Each group consists of $\frac{1}{2}$ of a pound of candy.

2. If we consider Frank and John as each representing one group, and David as representing half of a group, then the $14 should be distributed equally among $2\frac{1}{2}$ groups. Therefore, this is a *how many in each group* division problem. Each group should get

$$14 \div 2\frac{1}{2} = 14 \div \frac{5}{2} = 14 \times \frac{2}{5} = \frac{28}{5} = 5\frac{3}{5}$$

dollars. Therefore Frank and John should each get $5.60 and David should get half of that, which is $2.80.

3. According to the *how many in each group?* interpretation, the problem is solved by $4.50 ÷ .15 because $4.50 is put into .15 of a group. So the meal cost $4.50 ÷ .15 = $30.

4. Each problem, except for the first and last, requires different arithmetic to solve it.

   (a) This is asking: $\frac{1}{4}$ equals what times $\frac{1}{3}$? We solve this by calculating $\frac{1}{4} \div \frac{1}{3}$, which is $\frac{3}{4}$. We can also think of this as a division problem with the *how many groups?* interpretation because we want to know how many $\frac{1}{3}$ of a cup are in $\frac{1}{4}$ of a cup. According to the meaning of division, this is $\frac{1}{4} \div \frac{1}{3}$.

   (b) This is asking: what is $\frac{1}{4}$ of $\frac{1}{3}$? We solve this by calculating $\frac{1}{4} \times \frac{1}{3} = \frac{1}{12}$.

   (c) This is asking: what is $\frac{1}{3} - \frac{1}{4}$? The answer is $\frac{1}{12}$ which happens to be the same answer as in part (b), but the arithmetic to solve it is different.

   (d) Since $\frac{1}{4}$ cup of water fills $\frac{1}{3}$ of a plastic container, the full container will hold 3 times as much water, or $3 \times \frac{1}{4} = \frac{3}{4}$ of a cup. We can also think of this as a division problem with the *how many in each group?* interpretation. $\frac{1}{4}$ cup of water is put into $\frac{1}{3}$ of a group. We want to know how much is in one group. According to the meaning of division it's $\frac{1}{4} \div \frac{1}{3}$, which again is equal to $\frac{3}{4}$.

5. A simple story problem for $1 \div \frac{5}{7}$ is "how many $\frac{5}{7}$ of a cup of water are in 1 cup of water?" Figure 5.17 shows 1 cup of water and shows $\frac{5}{7}$ of a cup of water shaded. The shaded portion is divided into 5 equal parts and the full cup is 7 of those parts. So the full cup is $\frac{7}{5}$ *of the shaded part*. Thus there are $\frac{7}{5}$ of $\frac{5}{7}$ of a cup of water in 1 cup of water, so $1 \div \frac{5}{7} = \frac{7}{5}$.

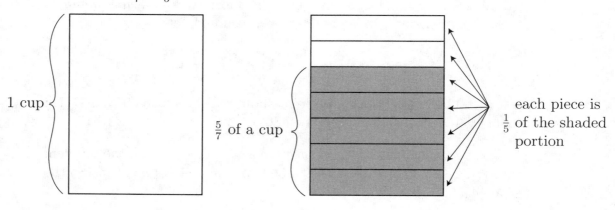

Figure 5.17: Solving $1 \div \frac{5}{7}$

6. (a) The number of $\frac{1}{2}$ years in $3\frac{3}{4}$ years is $3\frac{3}{4} \div \frac{1}{2}$. There will be that many groups of 4500 miles driven. So after $3\frac{3}{4}$ years you will have driven

$$
\begin{aligned}
(3\frac{3}{4} \div \frac{1}{2}) \times 4500 &= (\frac{15}{4} \div \frac{1}{2}) \times 4500 \\
&= \frac{15}{2} \times 4500 \\
&= 33,750
\end{aligned}
$$

miles.

(b) Since one year is 52 weeks there are $52 \div 6\frac{1}{2}$ groups of $6\frac{1}{2}$ weeks in a year. Mo will use 128 ounces for each of those groups, so Mo will use

$$
\begin{aligned}
(52 \div 6\frac{1}{2}) \times 128 &= (52 \div \frac{13}{2}) \times 128 \\
&= \frac{104}{13} \times 128 \\
&= 1,024
\end{aligned}
$$

ounces of detergent in a year.

(c) There are $32 \div 2\frac{1}{2}$ groups of $2\frac{1}{2}$ ounces in 32 ounces. Each of those groups makes 1 gallon. So the bottle makes $32 \div 2\frac{1}{2} = 12\frac{4}{5}$ gallons of weed killer.

7. One way to solve the problem is to determine how many $\frac{2}{3}$ units are in $\frac{5}{2}$ units. This will tell us how many of the $\frac{2}{3}$ unit long segments to lay end to end in order to get the $\frac{5}{2}$ unit long segment. Since $\frac{5}{2} \div \frac{2}{3} = \frac{15}{4} = 3\frac{3}{4}$, therefore there are $3\frac{3}{4}$ segments of length $\frac{2}{3}$ units in a segment of length $\frac{5}{2}$ units. So you need to form a line segment that is 3 times as long as the one pictured (above), plus another $\frac{3}{4}$ as long:

Problems for Section 5.5 on Dividing Fractions

1. Jose and Mark are making cookies for a bake sale. Their recipe calls for $2\frac{1}{4}$ cups of flour for each batch. They have 5 cups of flour. Jose and Mark realize that they can make two batches of cookies and that there will be some flour left. Since the recipe doesn't call for eggs, and since they have plenty of the other ingredients on hand, they figure they can make a fraction of a batch in addition to the two whole batches. Now Jose and Mark have a difference of opinion. Jose says that

$$5 \div 2\frac{1}{4} = 2\frac{2}{9}$$

and so he says that they can make $2\frac{2}{9}$ batches of cookies. But Mark says that two batches of cookies will use up $4\frac{1}{2}$ cups of flour, leaving $\frac{1}{2}$ left, so they should be able to make $2\frac{1}{2}$ batches. Mark draws the picture in Figure 5.18 to explain his thinking to Jose. Discuss the boys'

Figure 5.18: Cups of Flour

mathematics: what's right, what's not right, and why? If anything is incorrect, how could you modify it to make it correct?

2. Marvin has 11 yards of cloth to makes costumes for a play. Each costume requires $1\frac{1}{2}$ yards of cloth.

   (a) Solve the following two problems:

   (b) How many costumes can Marvin make and how much cloth will be left over?

   (c) What is $11 \div 1\frac{1}{2}$?

   (d) Compare and contrast your answers in part (a).

3. Suppose you are talking to someone who knows about fractions but does not know how to multiply or divide them. Explain to this person what $\frac{4}{3} \div \frac{1}{2}$ means, and show the person a way to determine the answer to this division problem with diagrams and logical thinking as opposed to a procedure.

4. Make up a story problem that can be solved by calculating $4 \div \frac{2}{3}$. Show how to solve the problem with pictures or diagrams. Explain clearly.

5. Write a story problem for $\frac{3}{4} \times \frac{1}{2}$ and another story problem for $\frac{3}{4} \div \frac{1}{2}$ (make clear which is which). In each case, use a diagram to help you solve the problem. Explain clearly how your diagrams show how to solve the problems.

6. Will has mowed $\frac{2}{3}$ of his lawn and so far it's taken him 45 minutes. How long will it take Will to mow the entire lawn (all together)?

   (a) Draw a picture and use the picture to help you solve the problem. Explain your reasoning.

   (b) Now solve the problem by interpreting it as a division problem. Which of the two interpretations of division does this use? Explain! (It might help you to first modify the problem so that it only involves whole numbers. In this case, you might want to think of Will as running a lawn service.)

7. Grandma's favorite muffin recipe uses $1\frac{3}{4}$ cups of flour for one batch of 12 muffins. Answer each of the following questions, *explaining clearly*

which of the two meanings of division you are using each time and why. In each case, calculate the answer without the use of a calculator.

    (a) How many cups of flour are in one muffin?

    (b) How many muffins does 1 cup of flour make?

    (c) If you have 3 cups of flour, then how many batches of muffins can you make? (Assume that you can make fractional batches of muffins.)

8. An article by Dina Tirosh, [53], discusses some common errors in division. The following problems are based on some of the findings of this article.

    (a) Tyrone says that $\frac{1}{2} \div 5$ doesn't make sense because 5 is bigger than $\frac{1}{2}$ and you can't divide a smaller number by a bigger number. Give Tyrone an example of a sensible story problem that is solved by $\frac{1}{2} \div 5$.

    (b) Kim says that $4 \div \frac{1}{3}$ can't be equal to 12 because when you divide, the answer is smaller. Kim thinks the answer should be $\frac{1}{12}$ because that is less than 4. Give Kim an example of a story problem that is solved by $4 \div \frac{1}{3}$ and explain why it makes sense that the answer really is 12.

9. In this section, we explained why the "invert and multiply" procedure for fraction division works in terms of the *how many groups?* interpretation of division. It is also possible to explain why the "invert and multiply" procedure works in terms of the *how many in each group?* interpretation of division. This problem will help you see how for the following division problem:

> Suppose $\frac{1}{2}$ of a bucket of blueberries fills $\frac{2}{3}$ of a plastic container. How many buckets of blueberries will it take to fill the whole plastic container?

    (a) Write the problem as a division problem.

    (b) Draw pictures to help you solve the problem.

    (c) Explain how your pictures help you see that $\frac{1}{2} \times \frac{3}{2}$ buckets of blueberries are needed to fill the whole plastic container. (*Actually,* your picture should show that $\frac{3}{2} \times \frac{1}{2}$ buckets of blueberries are needed, but multiplication is commutative.)

10. In this section, we explained why the "invert and multiply" procedure for fraction division works in terms of the *how many groups?* interpretation of division. It is also possible to explain why the "invert and multiply" procedure works in terms of the *how many in each group?* interpretation of division. This problem will help you see how for the following division problem:

> A new road is being built. The portion of the road that has been built is $\frac{2}{3}$ of a mile long and is $\frac{3}{4}$ of the length that the completed road will be. How long will the full road be when it is completed?

(a) Write the problem as a division problem.

(b) Draw pictures to help you solve the problem.

(c) Explain how your pictures help you see that the full road will be $\frac{2}{3} \times \frac{4}{3}$ miles long. (*Actually*, your picture should show that the full road will be $\frac{4}{3} \times \frac{2}{3}$ miles long, but multiplication is commutative.)

5.6   Dividing Decimals

When we divide (finite) decimals, such as

$$2.35\overline{)3.714}$$

the standard procedure is to move the decimal point in both the divisor (2.35) and the dividend (3.714) the same number of places to the right until the divisor becomes a whole number:

$$235\overline{)371.4}$$

Then a decimal point is placed above the decimal point in the (new) dividend (371.4):

$$235\overline{)371.4}^{\,\cdot}$$

From then on, division proceeds just as if one were doing the whole number division problem $3714 \div 235$, except that a decimal point is left in place for the answer. Why does this procedure of shifting the decimal points make sense?

One way to explain this is that by shifting the decimal points in both the divisor and the dividend, you have *multiplied and divided by the same power of 10*, thereby replacing the original problem with an equivalent problem that can be solved by previous methods. Recall that multiplying a decimal by a power of 10 shifts the digits in the decimal, and therefore can be thought of as shifting the decimal point. In our example,

$$235 = 2.35 \times 10^2,$$
$$371.4 = 3.714 \times 10^2$$

Therefore

$$371.4 \div 235 = (3.714 \times 10^2) \div (2.35 \times 10^2)$$
$$= 3.714 \div 2.35$$

Since we multiplied and divided by the same number, namely 10^2, therefore the original problem $3.714 \div 2.35$ is equivalent to the new problem $371.4 \div 235$. This is particularly nice when we re-write the division problems in fraction form:

$$\frac{371.4}{235} = \frac{3.714 \times 10^2}{2.35 \times 10^2}$$
$$= \frac{3.714}{2.35}$$

So the shifting of decimal points described above is really just a way to *replace the problem*

$$3.714 \div 2.35$$

with the equivalent problem

$$371.4 \div 235.$$

Class Activity 5L: Quick Tricks for Some Decimal Division Problems

Class Activity 5M: Understanding Decimal Division

Exercises for Section 5.6 on Dividing Decimals

1. Liz looked at the problem $.11 \div .125$ and almost immediately gave the answer .88, without using a calculator or paper and pencil. How might Liz have solved this problem so quickly?

2. Give two different explanations for why the two division problems

$$.29\overline{)1.7} \quad \text{and} \quad 29\overline{)170}$$

   have the same answer.

3. How are the two division problems

$$(5.7 \times 10^{27}) \div (3.1 \times 10^{15}) \quad \text{and} \quad 5.7 \div 3.1$$

   related?

Answers to Exercises for Section 5.6 on Dividing Decimals

1. Liz probably knows that $.125 = \frac{1}{8}$ and that to divide by $\frac{1}{8}$ you multiply by 8. It's easy to see that $.11 \times 8 = .88$.

2. Since

$$29 = .29 \times 10^2$$
$$\text{and}$$
$$170 = 1.7 \times 10^2$$
$$\text{therefore}$$
$$170 \div 29 = (1.7 \times 10^2) \div (.29 \times 10^2)$$
$$= 1.7 \div .29$$

because we just multiplied and divided by 10^2.

Another way to explain why the two division problem have the same answer is by interpreting Figure 5.19 in two different ways. On the one hand, if the smallest square shown in the diagram is interpreted as representing $\frac{1}{100}$, then the question in the diagram asks "how many .29 are in 1.7?" which means $1.7 \div .29 = ?$. On the other hand, if the smallest square is interpreted as representing 1, then the question in the diagram asks "how many 29 are in 170?" which means $170 \div 29 = ?$. Because the answer is the same either way, therefore the two division problems have the same answer. (From the diagram, you can see that the answer is between 5 and 6 because .29 is almost 30 little squares.)

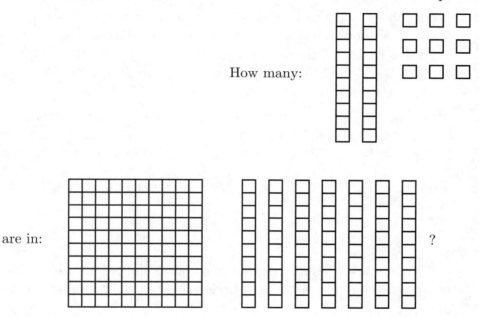

Figure 5.19: A Division Problem

3. You can get from $5.7 \div 3.1$ to $(5.7 \times 10^{27}) \div (3.1 \times 10^{15})$ by multiplying $5.7 \div 3.1$ by 10^{27} and dividing it by 10^{15}. A nice way to see this is with fractions:

$$\frac{5.7}{3.1} \times \frac{10^{27}}{10^{15}} = \frac{5.7 \times 10^{27}}{3.1 \times 10^{15}}.$$

But if you multiply by 10^{27} and divide by 10^{15}, then you've multiplied by 27 tens and divided by 15 tens, so the net effect is to multiply by $27 - 15 = 12$ tens. So $(5.7 \times 10^{27}) \div (3.1 \times 10^{15})$ is 10^{12} times as much as $5.7 \div 3.1$. Said symbolically:

$$(5.7 \times 10^{27}) \div (3.1 \times 10^{15}) = (5.7 \div 3.1) \times 10^{12}.$$

Problems for Section 5.6 on Dividing Decimals

1. Bruce has to do the division problem $75.23 \div 16.9$ longhand, but he can't remember what to do about decimal points. Instead, Bruce solves the division problem $7523 \div 169$ longhand and gets the answer 44.51. Bruce knows that the answer to $75.23 \div 16.9$ will be 44.51, except that the decimal point in 44.51 must be shifted over somehow.

   (a) Explain to Bruce how he can *reason about the sizes* of the numbers to determine where the decimal point is actually supposed to go in 44.51. *Do not* just tell Bruce the procedure for where to put the decimal point, and do not just repeat other methods of explanation from this section.

   (b) Show Bruce how to write 75.23 and 16.9 as *improper fractions* and use fraction division to determine where the decimal point should go in 44.51. *Make use of the work that Bruce has already done.*

2. In ordinary language, the term "divide" means "partition and make smaller," as in:

   Divide and conquer.

   In mathematics, does dividing always make smaller? In other words, if you start with a number N and divide it by another number M, is the resulting quotient $N \div M$ less than N?

   (a) Investigate this question by working out a number of examples. M and N can be *any* kinds of numbers: they can be fractions, integers, or decimals, they can be positive or negative.

   (b) Formulate a general answer to the following question: given a number N, for which numbers M is $N \div M$ less than N? (Your answer should involve cases, depending on the nature of N.)

3. Suppose you need to know how many thirtysecondths ($\frac{1}{32}$) of an inch .685 of an inch is (rounded to the nearest thirtysecondth of an inch). Explain why you can solve this problem by finding $.685 \times 32 = 21.92$ and rounding the 21.92 to the nearest whole number, namely 22. Then .685 of an inch, rounded to the nearest thirtysecondth of an inch, is $\frac{22}{32}$ of an inch. Don't just show that this method gives the right answer,

explain *why* it must give the right answer (so that you can see why it will work for other numbers too!).

4. Sarah is building a carefully crafted cabinet and calculates that she needs to cut a certain piece of wood 33.33 inches long. Sarah has a standard tape measure that shows subdivisions of one sixteenth ($\frac{1}{16}$) of an inch. How should Sarah measure 33.33 inches with her tape measure, using the closest sixteenth of an inch (how many whole inches and how many sixteenths of an inch)? Explain your reasoning.

5. The following information about two different snack foods is taken from their packages:

snack	serving size	calories in one serving	total fat in one serving
small crackers	28 g	140 calories	6 g
chocolate hearts	40 g	220 calories	13 g

(a) Explain how to interpret the information above so that the chocolate hearts have 117% more fat than the small crackers.

(b) Explain how to interpret the information above so that the chocolate hearts have 52% more fat than the small crackers.

(c) Explain how to interpret the information above so that the chocolate hearts have 38% more fat than the small crackers.

5.7   Ratio and Proportion

The concepts of ratio and proportion are traditionally considered some of the hardest in elementary mathematics. But ratios are essentially just fractions, and understanding and working with ratios and proportions really just involves understanding and working with multiplication, division, and fractions.

A **ratio** describes a certain kind of relationship between two quantities. If the ratio of flour to milk in a muffin recipe is 7 to 2, then that means that *for every 7 cups of flour you use, you must use 2 cups of milk.* Equivalently, for every 2 cups of milk you use, you must use 7 cups of flour. In general, to say that two quantities are in a ratio of A to B means that for every A units of the first quantity there are B units of the second quantity. Equivalently, for every B units of the second quantity there are A units of the first quantity.

To indicate a ratio we can use words, as in

the ratio of flour to milk is 7 to 2,

we can use a colon, as in

the ratio of flour to milk is 7 : 2,

or we can use a fraction, as in

the ratio of flour to milk is 7/2 (or $\frac{7}{2}$).

Later in this section we will see why it makes sense to write ratios as fractions.

A **proportion** is a statement that two ratios are equal. For example, if the ratio of flour to milk in a muffin recipe is 7 to 2, and if the muffin recipe is to use A cups of flour (an as yet unknown amount) and 3 cups of milk, then we have the proportion

$$\frac{A}{3} = \frac{7}{2},$$

which states that the two ratios A to 3 and 7 to 2 are equal.

Class Activity 5N: Mixtures and Ratios

Understanding Ratios and Proportions in Terms of Multiplication and Division

We will now see how to work with proportions simply by appealing to the meanings of multiplication and division. Consider various muffin recipes in which the ratio of flour to milk is 7 to 2, as shown in Figure 5.20. If the ratio of flour to milk in a muffin recipe is 7 to 2, then if we make a giant batch of muffins using 14 cups of flour, which is 2 groups of 7 cups, then we must use 2 groups of 2 cups of milk, which is 4 cups of milk. If we use 21 cups of flour, which is 3 groups of 7 cups, then we must use 3 groups of 2 cups of milk, which is 6 cups of milk. If we use $3\frac{1}{2}$ cups of flour, which is $\frac{1}{2}$ a group of 7 cups, then we must use $\frac{1}{2}$ a group of 2 cups of milk, which is 1 cup of milk. *Notice that to make sense of these calculations with proportions we only need to use the meanings of multiplication and division.*

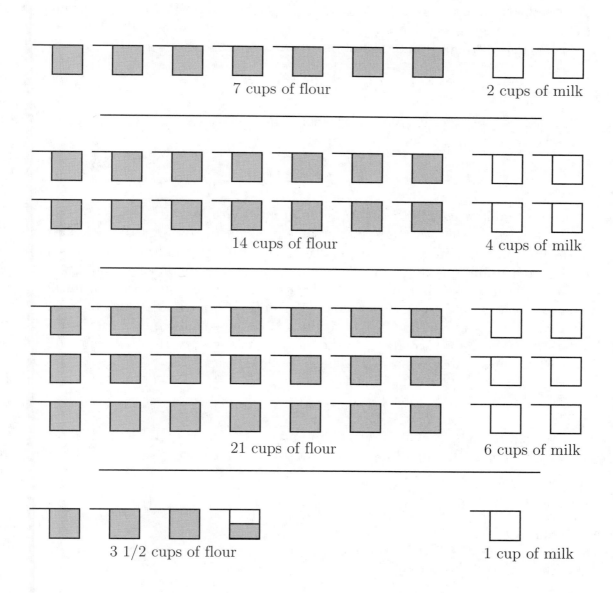

7 cups of flour

2 cups of milk

14 cups of flour

4 cups of milk

21 cups of flour

6 cups of milk

3 1/2 cups of flour

1 cup of milk

Figure 5.20: The Ratio 7 to 2 of Flour to Milk

So if you want to use 4 cups of flour in a muffin recipe in which the ratio of flour to milk is 7 to 2, how much milk should you use? First, determine how many groups of 7 are in 4. *According to the meaning of division* (the "how many groups?" interpretation) there are

$$4 \div 7 = \frac{4}{7}$$

groups of 7 in 4. Next, for each whole group of 7, we must use 2 cups of milk, so for $\frac{4}{7}$ of a group of 7, we use

$$\frac{4}{7} \cdot 2 = \frac{8}{7} = 1\frac{1}{7}$$

cups of milk, *according to the meaning of multiplication*.

Here is another way to solve the same problem: if you want to use 4 cups of flour in a muffin recipe in which the ratio of flour to milk is 7 to 2, then how much milk should you use? Again, we will rely only on the meanings of multiplication and division to solve this problem. We can first ask how much milk we will need for 1 cup of flour. If 7 cups of flour take 2 cups of milk, then we can think of the 2 cups of milk as divided equally among the 7 cups of flour. Therefore according to the meaning of division (the "how many in each group?" interpretation), each cup of flour takes

$$2 \div 7 = \frac{2}{7}$$

of a cup of milk. Now if we want to use 4 cups of flour, then we need to use 4 groups of $\frac{2}{7}$ of a cup of milk, in other words, we need to use

$$4 \cdot \frac{2}{7} = \frac{8}{7} = 1\frac{1}{7}$$

cups of milk.

The moral is that we can make sense of ratios and proportions by using the meanings of multiplication and division.

Ratios and Fractions

Ratios and fractions are very closely related. In fact, ratios give rise to fractions, and fractions give rise to ratios. For this reason, ratios are sometimes written as fractions. We will now see why this makes sense.

Consider a bread recipe in which the ratio of flour to water is 14 to 5, so that if we use 14 cups of flour, we will need 5 cups of water. If we think of dividing those 14 cups of flour equally among the 5 cups of water, then according to the meaning of division (the "how many in each group?" interpretation), each cup of water goes with

$$14 \div 5 = \frac{14}{5}$$

cups of flour. So in this recipe, *there are $\frac{14}{5}$ cups of flour per cup of water*. In this way, the ratio 14 to 5 of flour to water is naturally associated with the fraction $\frac{14}{5}$, which tells us *the number of cups of flour per cup of water*.

In general, a ratio A to B relating two quantities gives rise to the fraction $\frac{A}{B}$, which tells us the number of units of the first quantity there is *per unit amount of the second quantity*. The one exception, which doesn't even occur in ordinary circumstances, is when B is zero—in that case we cannot form the fraction $\frac{A}{B}$.

Similarly, a fraction (or other number) that tells us how much of one quantity there is per unit amount of another quantity, gives rise to a ratio between the two quantities. If we have a beverage recipe in which we use $\frac{2}{3}$ of a cup of cola per cup of lemon soda, then in this recipe, the ratio of cola to lemon soda is

$$\frac{2}{3} \quad \text{to} \quad 1.$$

Notice that if we want to use 3 cups of lemon soda then we should use 3 times as much cola, or

$$3 \cdot \frac{2}{3} = 2$$

cups of cola. Therefore we can also express the ratio of cola to lemon soda as

$$2 \quad \text{to} \quad 3.$$

In general, if there is $\frac{A}{B}$ of one quantity per unit amount of a second quantity, then the ratio of the first quantity to the second is

$$\frac{A}{B} \quad \text{to} \quad 1,$$

which is the same ratio as

$$B \cdot \frac{A}{B} \quad \text{to} \quad B \cdot 1,$$

(because this is using B times as much of both quantities) which is

$$A \text{ to } B.$$

In this way fractions give rise to ratios.

In general, a ratio

$$A \;:\; B$$

is equal to the ratio

$$N \cdot A \;:\; N \cdot B.$$

Think about recipes to see why this makes sense: suppose we have recipe in which the ratio of two ingredients is A to B. Then if we want to use $2 \cdot A$ of the first ingredient, we should use $2 \cdot B$ of the second ingredient; if we want to use $3 \cdot A$ of the first ingredient, we should use $3 \cdot B$ of the second ingredient; and so on—if we want to use $N \cdot A$ of the first ingredient, then we should use $N \cdot B$ of the second ingredient. This is exactly the same phenomenon as for fractions:

$$\frac{A}{B} = \frac{N \cdot A}{N \cdot B}.$$

The Logic Behind Solving Proportions by Cross-Multiplying

You are probably familiar with the technique of solving proportions by cross-multiplying. Why is this a valid technique for solving proportions? We will examine this now.

Consider a recipe for a bubbly fruit juice drink consisting of $\frac{1}{4}$ cup frozen fruit juice concentrate and 4 cups seltzer water. How much juice concentrate will you need if you want to use 6 cups seltzer water, and if you are using the same ratio of juice concentrate to seltzer water? A common method for solving such a problem is to set up the following proportion in fraction form,

$$\frac{\frac{1}{4}}{4} = \frac{A}{6},$$

where A represents the as yet unknown amount of juice concentrate you will need for 6 cups of seltzer water. We then "cross-multiply":

$$6 \cdot \frac{1}{4} = 4 \cdot A,$$

and therefore

$$A = \left(6 \cdot \frac{1}{4}\right) \div 4 = 1\frac{1}{2} \div 4 = \frac{5}{8},$$

so that you must use $\frac{5}{8}$ cups of juice concentrate for 6 cups of seltzer water.

Let's analyze the steps above. First, when we set the two fractions

$$\frac{\frac{1}{4}}{4}$$

and

$$\frac{A}{6}$$

equal to each other, why can we do that and what does that mean? If we think of the fractions as representing division, i.e.,

$$\frac{1}{4} \div 4$$

and

$$A \div 6,$$

then each of these expressions stands for the number of cups of juice concentrate per cup of seltzer water. We want to use the same amount of juice concentrate per cup of seltzer water either way, so therefore the two fractions should be equal to each other:

$$\frac{\frac{1}{4}}{4} = \frac{A}{6}.$$

Next, why do we "cross-multiply"? The reason we can cross-muliply is because two fractions are equal exactly when their "cross-multiples" are equal. Recall that the method of cross-multiplying for fractions is really just a shortcut for giving the two fractions a common denominator. We can see the same phenomenon even when we go back to the original ratios. The ratio

$$\frac{1}{4} \text{ to } 4$$

of juice concentrate to seltzer water is the same as the ratio

$$6 \cdot \frac{1}{4} \text{ to } 6 \cdot 4.$$

The ratio

$$A \text{ to } 6$$

of juice concentrate to water is the same as the ratio

$$4 \cdot A \quad \text{to} \quad 4 \cdot 6.$$

Both of the ratios

$$6 \cdot \frac{1}{4} \quad \text{to} \quad 6 \cdot 4$$

and

$$4 \cdot A \quad \text{to} \quad 4 \cdot 6$$

refer to 24 cups of seltzer water ($6 \cdot 4$ and $4 \cdot 6$), so therefore they must refer to the same amount of juice concentrate, which is $6 \cdot \frac{1}{4}$ cups on the one hand, and $4 \cdot A$ cups on the other hand. Therefore

$$6 \cdot \frac{1}{4} = 4 \cdot A.$$

Class Activity 5O: Solving Proportion Problems in a Variety of Ways

Class Activity 5P: Can You Always Use a Proportion?

Using Proportions: The Consumer Price Index

At some point you have probably heard older people wistfully recall the lower prices of days gone by, as in, "when I was young, a candy bar only cost a nickel!". As time goes by, most items become more expensive, due to inflation (computers are a notable exception). So in years past, one dollar bought more than a dollar buys now. But in years past most people also earned less than they do now. So what is a fair way to compare costs of items across years? What is a fair way to compare wages earned in different years? The standard way to make such comparisons is with the Consumer Price Index, which is generated by the Bureau of Labor Statistics of the U.S. Department of Labor. To use the Consumer Price Index, we must work with proportions.

According to the Bureau of Labor Statistics, the Consumer Price Index (CPI) is a measure of the average change over time in the prices paid by urban consumers for a market basket of consumer goods and services. The CPI market basket is developed from detailed expenditure information provided by families and individuals on what they actually bought. The CPI measures inflation as experienced by consumers in their day-to-day living expenses. This, and other information about the CPI is available on the Bureau of Labor Statistics website at

`http://www.bls.gov/cpi/home.htm`

Here is an example of how to use the CPI to compare salaries in different years. According to the Bureau of Labor Statistics, the average annual wages for elementary teachers in the U.S. was $38,600 in 1998 and $41,980 in the year 2000. Now the average wage was higher in 2000 than it was in 1998, but most items were also more expensive in 2000 than they were in 1998. So in 2000 did elementary teachers have more "buying power" than they did in 1998 or not? In other words, did elementary teachers' salaries go up faster than inflation or not? We can use the CPI to determine the answer to this question.

The CPI assigns a number to each year (in recent history). The table below shows the CPI for some selected years.

The Consumer Price Index selected years	
year	CPI
1940	14.0
1950	24.1
1960	29.6
1970	38.8
1980	82.4
1990	130.7
1991	136.2
1992	140.3
1993	144.5
1994	148.2
1995	152.4
1996	156.9
1997	160.5
1998	163.0
1999	166.6
2000	172.2

We interpret these numbers as follows. Because the CPI was 163.0 in 1998, and 172.2 in 2000, this means that for every $163.00 you would have spent toward a "market basket of consumer goods and services" in 1998, you would have had to spend $172.20 in the year 2000. We can use this to compare the teachers' salaries in 1998 and 2000. Think of each group of $163.00 as buying a "unit" of a market basket of goods and services in 1998. So in 1998, the average elementary teacher's salary could buy

$$38,600 \div 163 = 236.81$$

units of a market basket of goods and services. To buy the same number of units of a market basket of goods and services in 2000, namely 236.81 units, you would have needed

$$236.81 \times \$172.2 = \$40,779$$

in 2000, because each unit of a market basket of goods and services cost $172.20 in 2000. In other words, $40,779 in 2000 had the same buying power as did $38,600 in 1998. Given that in the year 2000 elementary teachers made an average of $41,980, which is more than $40,779, therefore elementary teachers had more buying power in 2000 than they did in 1998. So from 1998 to 2000, elementary teachers' salaries went up faster than inflation.

Here is another way to compare elementary teachers average salaries in 1998 and 2000 using the CPI. Because the CPI was 163.0 in 1998, and 172.2 in 2000, we can again think of $163.00 in 1998 and $172.20 in 2000 as buying a "unit" of a market basket of consumer goods and services. Now since

$$172.20 \div 163.00 = 1.05644,$$

therefore a "unit" of a market basket of goods and services cost 1.056 times as much in 2000 as it did in 1998. Therefore a salary of $38,600 in 1998 had the same buying power as did a salary 1.056 times as much, namely

$$1.056 \times \$38,600 = \$40,779$$

did in 2000. We reach the same conclusion as in the previous paragraph.

Either of the calculations of the previous two paragraphs are called **adjusting for inflation**. In both paragraphs, we converted $38,600, the average salary for elementary teachers in 1998, to *2000 dollars*, in which it became $40,779. So a 1998 salary of $38,600 had the same buying power as a salary of $40,779 did in 2000. Notice that another way to set up the calculations for inflation adjusting would be to set up either of the following two proportions:

$$\frac{38,600}{163} = \frac{x}{172.2},$$

or

$$\frac{172.2}{163.0} = \frac{x}{38,600},$$

where x stands for the dollar amount in 2000 that had the same buying power as did $38,600 in 1998.

Class Activity 5Q: The Consumer Price Index

Exercises for Section 5.7 on Ratio and Proportion

1. Suppose that a logging crew can cut down five acres of trees every two days. Assume that the crew works at a steady rate. Solve the following problems using logical thinking and using the most elementary reasoning you can. Explain your reasoning clearly.

   (a) How many days will it take the crew to cut 8 acres of trees? Give your answer as a mixed number.

   (b) Now suppose there are three logging crews that all work at the same rate as the one above. How long will it take these three crews to cut down 10 acres of trees?

2. In order to reconstitute a medicine properly, a pharmacist must mix 10 milliliters (ml) of a liquid medicine for every 12 ml of water. If one dose of the medicine/water mixture must contain $2\frac{1}{2}$ ml of *medicine*, then how many milliliters of *medicine/water* mixture provides one dose of the medicine? Solve this problem using logical thinking and using the most elementary reasoning you can. Explain your reasoning clearly.

3. If 3 people take 2 days to paint 5 fences, how long will it take 2 people to paint 1 fence? (Assume that the fences are all the same size and the painters are all equally good workers, and work at a steady rate.) Can this problem be solved by setting up the following proportion to find

how long it will take 2 people to paint 5 fences?

$$\frac{3 \text{ people}}{2 \text{ days}} = \frac{2 \text{ people}}{x \text{ days}}$$

Solve the problem by thinking logically about the situation. Explain your reasoning clearly.

4. If a candy bar cost a nickel in 1960, then how much would you expect to have to pay for the same size and type of candy bar in 2000, according to the CPI?

5. (a) What amount of money had the same buying power in 1950 as $50,000 did in 2000?

   (b) What amount of money had the same buying power in 1980 as $50,000 did in 2000?

6. If an item cost $5.00 in 1990 and $6.50 in 2000, did its price go up faster, slower or the same as inflation?

7. Suppose that from 1990 to 2000 the price of a 16 ounce box of brand A cereal went from $2.39 to $3.89. *After adjusting for inflation*, determine by what percent the price of Brand A cereal went up or down from 1990 to 2000?

Answers to Exercises for Section 5.7 on Ratio and Proportion

1. (a) Because the crew cuts 5 acres every 2 days, it will cut half as much in 1 day, namely $2\frac{1}{2}$ acres. In 8 days it will cut 8 times as much, namely $8 \div 2\frac{1}{2} = 3\frac{1}{5}$ days.

   (b) In part (a) we saw that one crew cuts $2\frac{1}{2}$ acres per day. Therefore three crews will cut 3 times as much per day, which is $3 \cdot 2\frac{1}{2} = 7\frac{1}{2}$ acres per day. To determine how many days it will take to cut 10 acres we must figure out how many groups of $7\frac{1}{2}$ are in 10. This is solved by the division problem $10 \div 7\frac{1}{2}$. Because $10 \div 7\frac{1}{2} = 10 \div \frac{15}{2} = \frac{10}{1} \cdot \frac{2}{15} = \frac{20}{15} = 1\frac{5}{15} = 1\frac{1}{3}$, therefore it will take the 3 crews $1\frac{1}{3}$ days to cut 10 acres.

2. Method 1: For 10 ml of medicine there are 12 ml of water. So if we think of dividing the 12 ml of water equally among the 10 ml of medicine, we see that there are $12 \div 10 = \frac{6}{5}$ ml of water for each milliliter of medicine. So for $2\frac{1}{2}$ ml of medicine there should be $2\frac{1}{2}$ times as much water, namely

$$2\frac{1}{2} \cdot \frac{6}{5} = \frac{5}{2} \cdot \frac{6}{5} = 3$$

milliliters of water. All together, the $2\frac{1}{2}$ ml of medicine and the 3 ml of water makes $5\frac{1}{2}$ ml of medicine/water mixture.

Method 2: For every group of 10 ml of medicine, the pharmacist mixes in a group of 12 ml of water. How many groups of 10 ml are in $2\frac{1}{2}$ ml? You might just see right away that $2\frac{1}{2}$ ml is $\frac{1}{4}$ of a group of 10 ml. Otherwise you can solve this by the division problem:

$$2\frac{1}{2} \div 10 = \frac{5}{2} \div \frac{10}{1} = \frac{5}{2} \cdot \frac{1}{10} = \frac{1}{4}.$$

So there is $\frac{1}{4}$ of a group of 10 in $2\frac{1}{2}$. Therefore the pharmacist should use $\frac{1}{4}$ of a group of 12 ml of water, which is 3 ml of water. All together, the $2\frac{1}{2}$ ml of medicine and the 3 ml of water makes $5\frac{1}{2}$ ml of medicine/water mixture.

3. The proportion

$$\frac{3 \text{ people}}{2 \text{ days}} = \frac{2 \text{ people}}{x \text{ days}}$$

is not valid for this situation because when more people are painting, it will take *less* time to paint a fence. It is not the case that for each group of 3 people there are it will take 2 days to paint a fence. Therefore the relationship between the number of people and the number of days it takes to paint a fence is not a ratio.

Thinking logically, if 3 people take 2 days to paint 5 fences, then those 3 people will take $2 \div 5 = \frac{2}{5}$ of a day to paint just one fence (dividing the 2 days equally among the 5 fences). If just one person was paiting, it would take 3 times as long to paint the fence, namely $3 \cdot \frac{2}{5} = \frac{6}{5}$ of a day (which is $1\frac{1}{6}$ days). With 2 people painting, it will take half as much time to paint the fence, namely $\frac{3}{5}$ of a day.

4. The CPI was 29.6 in 1960 and 172.2 in 2000, therefore $.05 in 1960 had the same buying power as did

$$\frac{172.2}{29.6} \cdot \$.05 = \$.29$$

in 2000.

5. (a) In 2000, the CPI was 172.2, so $50,000 could buy $50,000 \div 172.2 = 290.36$ "units" of a market basket of goods and services. In 1950 the CPI was 24.1, so the same 290.36 units of a market basket of goods and services cost $290.36 \times \$24.10 = \$6998$ in 1950. Therefore $6998 in 1950 had the same buying power as did $50,000 in 2000.

   (b) In 2000, the CPI was 172.2; in 1980, the CPI was 82.4. Therefore a "unit" of a market basket of goods and services cost $82.4 \div 172.2 = .47851$ times as much in 1980 as it did in 2000. Therefore $50,000 in 2000 had the same buying power as did $.47851 \times \$50,000 = \$23,926$ in 1980.

6. According to the CPI, $5.00 in 1990 had the same buying power as did

$$\frac{172.2}{130.7} \cdot \$5.00 = \$6.59$$

in 2000. So if an item cost $5.00 in 1990 and $6.50 in 2000, then its price went up a little slower than inflation.

7. In 2000, the CPI was 172.2 and in 1990, the CPI was 130.7, therefore, adjusting for inflation, a price of $2.39 in 1990 corresponds to

$$\frac{172.2}{130.7} \cdot \$2.39 = \$3.15$$

in 2000. Therefore, to make a fair comparison between the prices, we should calculate the percent increase from $3.15 to $3.89, instead of the percent increase from $2.39 to $3.89. Because

$$\frac{3.89}{3.15} = 1.23,$$

therefore the price of the cereal increased by 23%, after adjusting for inflation.

Problems for Section 5.7 on Ratio and Proportion

1. You can make grape juice by mixing 1 can of frozen grape juice concentrate with 3 cans of water. What are other ways of mixing grape juice concentrate with water so that the result will taste the same? Explain how this problem is related to ratios and proportions.

2. Which mixture will have a stronger grape flavor?

   • A mixture of 2 parts grape juice concentrate to 5 parts water, or

   • a mixture of 4 parts grape juice concentrate to 7 parts water?

   Explain your reasoning clearly.

3. A recipe that serves 6 people calls for $1\frac{1}{2}$ cups of rice. How much rice will you need to serve 8 people (assuming that the ratio of people to cups of rice stays the same)?

   (a) Solve this problem with a proportion. Explain carefully why this method solves the problem correctly. Be sure to address each step in the method (including the initial step of setting up the proportion—be sure to explain what the two fractions mean and why you set them equal to each other).

   (b) Now solve the same problem in a different way by using logical thinking and by using the most elementary reasoning you can. Explain your reasoning clearly.

4. A 5 gallon bucket filled with water is being pulled from the ground up to a height of 20 feet at a rate of 2 feet every 15 seconds. The bucket has a hole in it, so that water leaks out of the bucket at a rate of 1 quart ($\frac{1}{4}$ gallon) every 3 minutes. How much of the water will be left in the bucket by the time the bucket gets to the top? Solve this problem by using logical thinking and by using the most elementary reasoning you can. Explain your reasoning clearly.

5. Suppose that you have two square garden plots: one is 10 feet by 10 feet and the other is 15 feet by 15 feet. You want to cover both gardens with a one inch layer of mulch. If the 10 by 10 garden took $3\frac{1}{2}$ bags

of mulch, could you figure out how many bags of mulch you'd need for the 15 by 15 garden by setting up a proportion:

$$\frac{3\frac{1}{2}}{10} = \frac{x}{15}?$$

Explain clearly why or why not. If the answer is no, is there another proportion that you could set up? It may help you to draw pictures of the gardens.

6. If 6 men take 3 days to dig 8 ditches, then how long would it take 4 men to dig 10 ditches? Assume that all the ditches are the same size and take equally long to dig, that all the men work at the same steady rate. Solve this problem by using logical thinking and by using the most elementary reasoning you can. Explain your reasoning clearly.

7. Suppose that 400 pounds of freshly picked tomatoes are 99% water by weight. After one day, the same tomatoes only weigh 200 pounds due to evaporation of water. (The tomatoes consist of water and solids. Only the water evaporates, the solids remain.)

   (a) How many pounds of solids are present in the tomatoes? (Notice that this is the same when they are freshly picked as after one day.)

   (b) Therefore, when the tomatoes weigh 200 pounds, what percent of the tomatoes is water?

   (c) Is it valid to use the following proportion to solve for the percent of water, x, in the tomatoes when they weigh 200 pounds?:

   $$\frac{.99}{400} = \frac{x}{200}$$

   If not, why not?

8. Josie made $40,000 in 1995 and $50,000 in 2000. In absolute dollars, that's a 25% increase in Josie's salary from 1995 to 2000. But what about in terms of buying power? After adjusting for inflation, how much higher or lower was Josie's salary in 2000 than it was in 1995? Give your answer in percent.

9. Suppose that a person earned $30,000 in 1995 and got a 3% raise every year after that until the year 2000 (so that every year, their salary was 3% higher than their previous year's salary). After adjusting for inflation, how much higher or lower was the person's salary in 2000 than it was in 1995? Give your answer in percent.

10. According to the University of Georgia Fact Books from 1976 and 1996, expenditures on instruction at UGA were 31.1% of the total budget in 1976 and 18.7% of the total budget in 1996. In 1976 UGA spent $42,672,636 on instruction (and departmental research), whereas in 1996, UGA spent $139,672,851 on instruction. The total enrollment in 1976 was 22,879, and in 1996 it was 29,404. How can we compare expenditures on instruction in 1976 and 1996? We can take several different approaches, for example:

    John says that expenditures on instruction went down by 12.4%.

    Alice says that expenditures on instruction went down by 39.9%.

    Richard says that expenditures on instruction went up by 227%.

    Tia says that Richard's number is misleading because it doesn't adjust for inflation.

    Beatrice says that the number of students should also be taken into account.

    (a) In what sense is John correct?

    (b) In what sense is Alice correct? (Compare the percentages—treat them as ordinary numbers, such as the prices of shoes. If a pair of shoes goes from $31.10 to $18.70, what percent decrease does that represent?)

    (c) In what sense is Richard correct?

    (d) Address Tia's objection. In 1976 the CPI was 56.9; in 1996 the CPI was 156.9. If you adjust for inflation and work in 1996 dollars, then by what percent did expenditures on instruction go up or down from 1976 to 1996?

    (e) Address Beatrice's point. Compare inflation adjusted expenditures on instruction per student in 1976 and 1996. By what percent did the per student expenditures on instruction go up or down?

Chapter 6

More About Numbers

In this chapter we will study some of the theory of numbers, including factors, multiples, prime numbers, even and odd numbers, and some quick tests for determining if a number is divisible by another number. Most elementary school curricula do not emphasize these topics, however some children do learn about many of these topics in the upper elementary grades. Even young children can, and often do, learn about even and odd numbers.

6.1   Factors and Multiples

In this brief section we study situations where counting numbers decompose into products of counting numbers.

If A and B are counting numbers and if $A \div B$ is a whole number with no remainder, then we say that A is **evenly divisible** (or **divisible**) by B, or that B **divides** A. For example, 12 is divisible by 4 and 5 divides 30.

If A and B are counting numbers, then we say that B is a **factor** or a **divisor** of A if there is a counting number C such that

$$A = B \times C.$$

So 7 is a factor of 21 because

$$21 = 7 \times 3,$$

The numbers $3, 1, 21$ are also a factors of 21, and these are the only other factors of 21. Notice that the concepts of *divisibility* and of *factor* are closely linked: *a counting number B is a factor of a counting number A exactly when A is divisible by B.*

If A and B are counting numbers, then we say that A is a **multiple** of B if there is a counting number C such that

$$A = C \times B.$$

So 44 is a multiple of 11 because

$$44 = 4 \times 11.$$

Similarly, the numbers 55 and 66 are also multiples of 11. The list of multiples of 11 is infinitely long:

$$11, 22, 33, 44, 55, 66, \ldots.$$

Notice that the concepts of *factors* and *multiples* are closely linked: a counting number A is a multiple of a counting number B exactly when B is a factor of A. For this reason it's easy to get the concepts of *factors* and *multiples* confused. Remember that the *factors* of a number are the numbers you get from writing the original number as a product, i.e., from "breaking down" the number by dividing it. On the other hand, the *multiples* of a number are the numbers you get by multiplying the number by counting numbers, i.e., by "building up" the number by multiplying it.

Above, the word *factor* was defined as a noun. However, the word *factor* can also be used as a verb. If A is a counting number, then **to factor** A means to write A as a product of two or more counting numbers, each of which is smaller than A. So we can factor 18 as 9 times 2:

$$18 = 9 \times 2.$$

But notice that we can factor 18 even further because we can factor 9 as 3 times 3, so:

$$18 = 3 \times 3 \times 2.$$

Class Activity 6A: Factors, Multiples, and Rectangles

Class Activity 6B: Writing Problems about Factors and Multiples

Class Activity 6C: Do Factors Always Come in Pairs?

Exercises for Section 6.1 on Factors and Multiples

1. What are the factors of the number 72?

2. What are the multiples of the number 72?

3. Tanya has a large bulletin board that is 54 inches wide across the top. Tanya wants to put a repeating pattern of scrolls (such as the pattern of scrolls shown in Figure 6.1) across the top of her bulletin board in such a way that the length of each scroll is a whole number of inches and a whole number of scrolls fit perfectly across the top of the bulletin board (i.e., there aren't any partial scrolls at the end). What are the options for how long Tanya can make each scroll? In each case, how

many scrolls will there be along the top of the bulletin board? Solve this scroll problem and write a related problem about factors or multiples (whichever is most appropriate).

Figure 6.1: A Repeating Pattern of Scrolls

4. Tawanda has painted many dry noodles and will put strings through them to make necklaces that are completely filled with noodles, as in Figure 6.2. Each noodle is 4 centimeters long. What are the lengths of the necklaces Tawanda can make? Solve this necklace problem and write a related problem about factors or multiples (whichever is most appropriate).

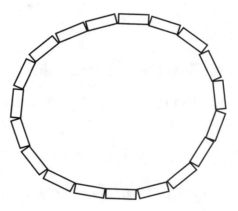

Figure 6.2: A Noodle Necklace

Answers to Exercises for Section 6.1 on Factors and Multiples

1. The factors of the number 72 are the counting numbers that divide 72 evenly. They are:

$$1, 2, 3, 4, 6, 8, 9, 72.$$

2. The multiples of 72 are the numbers that can be expressed as 72 times another counting number. They are:

$$72, 144, 216, 288, \ldots$$

because these numbers are

$$72 \times 1, 72 \times 2, 72 \times 3, 72 \times 4, \ldots.$$

3. Because the number of scrolls times the length of each scroll in inches is equal to 54 inches, therefore this problem is basically the same as finding all the ways to factor 54 into a product of two whole numbers. The ways of factoring 54 into a product of two whole numbers are:

$$1 \times 54, \quad 2 \times 27, \quad 3 \times 18, \quad 6 \times 9,$$

and the reverse factorizations:

$$54 \times 1, \quad 27 \times 2, \quad 18 \times 3, \quad 9 \times 6.$$

So Tanya can use 1 scroll that is 54 inches long, or 2 scrolls that are 27 inches long, or 3 scrolls that are 18 inches long, or 6 scrolls that are 9 inches long, or 9 scrolls that are 6 inches long, or 18 scrolls that are 3 inches long, or 27 scrolls that are 2 inches long, or 54 scrolls that are 1 inch long.

4. Whenever Tawanda makes a necklace from these noodles, the length of the necklace will be the number of noodles he used, times 4 centimeters. Therefore this problem is basically a problem about finding the multiples of 4, except of course that some of the multiples of 4 would make necklaces that are too short to be practical, and some multiples of 4 would make necklaces that Tawanda couldn't make because he doesn't have enough noodles. The multiples of 4 are

$$4, 8, 12, 16, 20, 24, \ldots$$

Problems for Section 6.1 on Factors and Multiples

1. Write a problem about a realistic situation involving concrete objects that involves the concept of factors. Explain how your problem involves the concept of factors.

2. Write a problem about a realistic situation involving concrete objects that involves the concept of multiples. Explain how your problem involves the concept of multiples.

3. Johnny says that 3 is a multiple of 6 because you can put 3 cookies into 6 groups by putting $\frac{1}{2}$ of a cookie in each group. Discuss Johnny's idea in detail: in what way does he have the right idea about what the term *multiple* means, and in what way does he need to modify his idea?

4. Solve the following two problems. Are the answers different or not? Explain why or why not.

   (a) Josh has 1159 bottle caps in his collection. In how many different ways can Josh break his bottle-cap collection into groups so that the same number of bottle caps are in each group and so that there are no bottle caps left over (i.e., not in a group)?

   (b) How many different rectangles can be made whose side lengths, in centimeters, are couting numbers and whose area is 1159 square centimeters?

6.2   Prime Numbers

In this section we will study the prime numbers, which can be considered the "building blocks" of the counting numbers.

The **prime numbers** are the counting numbers other than 1 that are divisible only by 1 and themselves. In other words, the prime numbers are the counting numbers that cannot be factored. For example, 2, 3, 5, and 7 are prime numbers, but 6 is not because $6 = 2 \times 3$.

The prime numbers are considered to be the *building blocks of the counting numbers*. Why? Because every whole number from 2 onward is either a prime number or can be factored as a product of prime numbers. For example:

$$
\begin{aligned}
145 &= 29 \times 5 \\
2009 &= 41 \times 7 \times 7 \\
264,264 &= 13 \times 11 \times 11 \times 7 \times 3 \times 2 \times 2 \times 2
\end{aligned}
$$

In this way, all natural numbers from 2 onward are "built" from prime numbers.

The prime numbers are to the natural numbers as atoms are to matter—basic and fundamental. And just as many physicists study atoms, so too, many mathematicians study the prime numbers. As with geometry, many fundamental contributions to knowledge about prime numbers were made by mathematicians of ancient Greece. Eratosthenes (275 – 195 B.C.) discovered a nice method for finding and listing prime numbers, called the *Sieve of Eratosthenes*.

Class Activity 6D: The Sieve of Eratosthenes

The Trial Division Method for Determining if a Number is Prime

How can you tell whether or not a number, such as 239, is a prime number? You could use the Sieve of Eratosthenes to find all prime numbers up to 239, but that would be slow. A faster way is to divide your number by 2, if it's not divisible by 2, divide it by 3; if it's not divisible by 3, divide it by 5; if it's not divisible by 5, divide it by 7, and so on, dividing by consecutive prime numbers. If you ever find a prime number that divides your number evenly, then your number is not prime; otherwise, it is prime.

This method for determining whether or not a number is prime is called the method of **trial division**.

Why is this method a valid way to determine if a number is prime, and how do you know when to stop trying to divide? The next class activity will examine these questions.

Class Activity 6E: The Trial Division Method for Determining if a Number is Prime

Factoring into Products of Prime Numbers

The prime numbers are "building blocks" of the counting numbers because every counting number can be broken into a product of prime numbers. How do we do this?

A convenient way to factor a counting number into a product of prime numbers is to create a **factor tree** as shown in Figure 6.3. Starting with a given counting number, such as 600, you may immediately see a way to

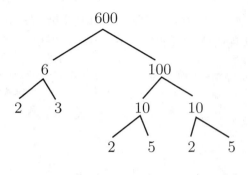

$$600 = 2 \times 3 \times 2 \times 5 \times 2 \times 5$$

Figure 6.3: A Factor Tree for 600

factor it, for example,

$$600 = 6 \times 100.$$

Record this factorization below the number 600, as shown in Figure 6.3. Then proceed by factoring 6 and 100 separately, and recording these factorizations below each number. Continue until all your factors are prime numbers. Then the numbers at the bottom of your factor tree are the prime factors of your original number. From Figure 6.3 we see that

$$600 = 2 \times 3 \times 2 \times 5 \times 2 \times 5.$$

You might want to rearrange your factorization by placing identical primes next to each other:

$$600 = 2 \times 2 \times 2 \times 3 \times 5 \times 5.$$

Using the notation of exponents, we can also write

$$600 = 2^3 \times 3 \times 5^2.$$

What if you don't immediately see a way to factor your number? For example, how can we factor 26741? In this case, try to divide it by consecutive prime numbers until you find a prime number that divides it evenly. We find that 26741 is not divisible by 2, by 3, by 5, or by 7, but it is divisible by 11:

$$26741 = 11 \times 2431.$$

Now we need to factor 2431. It can't be divisible by 2, by 3, by 5, or by 7 (because if it were, 26741 would also be divisible by one of these). But we must check if 2431 is divisible by 11, and sure enough, it is:

$$2431 = 11 \times 221.$$

Now check if 221 is divisible by 11. It isn't. So check if 221 is divisible by the next prime, 13. It is:

$$221 = 13 \times 17.$$

Collecting up the prime factors we have found, we conclude that

$$26741 = 11 \times 11 \times 13 \times 17.$$

The factor tree in Figure 6.4 records how we factored 26741.

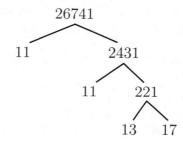

$$26741 = 11 \times 11 \times 13 \times 17$$

Figure 6.4: A Factor Tree for 26741

Although we will not prove it, it can be proven that every counting number other than 1 is either a prime number or can be factored into a product of prime numbers. Furthermore, no matter how you carry out the factorization of a counting number, you will always get the exact same prime factors, although perhaps in a different order. For example, you might factor the number 31603 as

$$31603 = 11 \times 13 \times 13 \times 17,$$

or perhaps as

$$31603 = 13 \times 11 \times 17 \times 13.$$

But you could never factor the number 31603 into a product of different prime numbers, say involving the prime 7 or the prime 19.

How Many Prime Numbers Are There?

Mathematicians through the ages have been fascinated by prime numbers. There are many interesting facts and questions about them. For instance, does the list of prime numbers go on forever, or does it ever come to a stop? Ponder this for a moment before you read on.

Do you have a definite opinion about this or are you unsure? If you do have a definite opinion, can you give convincing evidence for it? This is not so easy!

It turns out that the list of prime numbers does go on forever. But how do we know this? It certainly isn't possible to demonstrate this by listing all the prime numbers. It is not at all obvious why there are infinitely many prime numbers, but there is a beautiful line of reasoning that proves this. It is a perfect little gem that shows the human spirit of invention as does Mozart's gorgeous *Ave Verum*, or a Shakespeare sonnet:

Let me not to the marriage of true minds
Admit impediments. Love is not love
Which alters when it alteration finds,
Or bends with the remover to remove:
O no! it is an ever-fixed mark
That looks on tempests and is never shaken;
It is the star to every wandering bark,
Whose worth's unknown, although his height be taken.
Love's not Time's fool, though rosy lips and cheeks
Within his bending sickle's compass come:
Love alters not with his brief hours and weeks,
But bears it out even to the edge of doom.
If this be error and upon me proved,
I never writ, nor no man ever loved.
—*William Shakespeare*

In about 300 B.C., Euclid, another mathematician who lived in ancient Greece, wrote the most famous mathematics book of all time, *The Elements*. In it appears an argument that proves there are infinitely many prime numbers [16, Proposition 20 of book IX]. The following is a variation of that argument, presented in more modern language.

Suppose that
$$P_1, P_2, P_3, \ldots, P_n$$

is a list of n prime numbers. Consider the number

$$P_1 \cdot P_2 \cdot P_3 \cdot \ldots \cdot P_n + 1,$$

which is just the product of the list of prime numbers with the number one added on. Then this new number is not divisible by any of the primes $P_1, P_2, P_3, \ldots, P_n$ on our list, because each of these leave a remainder of 1 when our new number is divided by any of these numbers. Therefore, either $P_1 \cdot P_2 \cdot P_3 \cdot \ldots \cdot P_n + 1$ is a prime number which is not on our list of primes or, when we factor our number $P_1 \cdot P_2 \cdot P_3 \cdot \ldots \cdot P_n + 1$ into a product of prime numbers, each of these prime factors is a new prime number that is not on our list. Therefore, no matter how many prime numbers we started with, we can always find new prime numbers, and therefore there must be infinitely many prime numbers.

Exercises for Section 6.2 on Prime Numbers

1. What is a prime number?

2. What is the Sieve of Eratosthenes and how does it work?

3. For each of the following numbers, determine if it is prime. If it is not prime, factor the number into a product of prime numbers.

   (a) 1081
   (b) 1087
   (c) 269059
   (d) 2081
   (e) 1147

4. Given that $550 = 2 \cdot 5 \cdot 5 \cdot 11$, find all the factors of of 550.

Answers to Exercises for Section 6.2 on Prime Numbers

1. See text.

2. See Class Activity 6D.

3. (a) $1081 = 23 \times 47$.

   (b) 1087 is prime.

   (c) $269059 = 7 \times 7 \times 17 \times 17 \times 19$.

   (d) 2081 is prime.

   (e) $1147 = 31 \times 37$.

4. 1 and 550 are factors of 550. Given any other factor of 550, when this factor is factored, it must be a product of some of the primes in the list 2, 5, 5, 11. (If not, you would get a different way to factor 550 into a product of prime numbers.) Therefore the factors of 550 are

$$1, \ 550, \ 2, \ 5, \ 11, \ 2 \cdot 5 = 10, \ 2 \cdot 11 = 22, 5 \cdot 5 = 25, \ 5 \cdot 11 = 55,$$

$$2 \cdot 5 \cdot 5 = 50, \ 2 \cdot 5 \cdot 11 = 110, \ 5 \cdot 5 \cdot 11 = 275.$$

Problems for Section 6.2 on Prime Numbers

1. For which counting numbers, N, is there only one rectangle whose side lengths, in inches, are counting numbers and whose area, in square inches, is N? Explain!

2. For each of the following numbers, determine if it is a prime number. If it is not a prime number, factor the number into a product of prime numbers.

   (a) 8303

   (b) 3719

   (c) 3721

   (d) $80,000$

3. Without calculating the number $19 \times 23 + 1$, explain why this number is not divisible by 19 or by 23.

4. Suppose you started with the list of primes 5, 7 in Euclid's proof that there are infinitely many primes. Then you would form the number $5 \times 7 + 1 = 36$, and you come up with the new prime numbers 2 and 3 because $36 = 2 \times 2 \times 3 \times 3$. Which new prime numbers would you come up with if your starting list was the following?

(a) 2, 3, 5

(b) 2, 3, 5, 7

(c) 3, 5

6.3 Greatest Common Factor and Least Common Multiple

In the last section, we discussed factors and multiples of individual numbers. If we have two or more counting numbers in mind, we can consider the factors that the two numbers have in common and the multiples that the two numbers have in common. This leads to the concepts of *greatest common factor* and *least common multiple*, which we will discuss briefly in this section.

If you have two or more counting numbers, then the **greatest common factor**, abbreviated **GCF**, or **greatest common divisor**, abbreviated **GCD**, of these numbers is the greatest counting number that is a factor of all the given counting numbers. For example, what is the GCF of 12 and 18? The factors of 12 are

$$1, \ 2, \ 3, \ 4, \ 6, \ 12$$

and the factors of 18 are

$$1, \ 2, \ 3, \ 6, \ 9, \ 18.$$

Therefore the common factors of 12 and 18 are the numbers that are common to the two lists, namely

$$1, \ 2, \ 3, \ 6.$$

Because the greatest of these numbers is 6, therefore the greatest common factor of 12 and 18 is 6.

Similarly, if you have two or more counting numbers, then the **least common multiple**, abbreviated **LCM**, of these numbers is the least counting number that is a multiple of all the given numbers. For example, what is the LCM of 6 and 8? The multiples of 6 are

$$6, \ 12, \ 18, \ 24, \ 30, \ 36, \ 42, \ 48, \ 54, \ 60, \ 66, \ 72, \ 78, \ldots,$$

and the multiples of 8 are

$$8, \ 16, \ 24, \ 32, \ 40, \ 48, \ 56, \ 64, \ 72, \ 80, \ldots.$$

Therefore the common multiples of 6 and 8 are the numbers that are common to the two lists, namely

$$24, \ 48, \ 72, \ldots$$

Because the least of these numbers is 24, therefore the least common multiple of 6 and 8 is 24.

   Although there are other methods for calculating GCFs and LCMs, we can always use the definition of these concepts to calculate GCDs and LCMs.

Class Activity 6F: Concrete Problems Involving Greatest Common Factors and Least Common Multiples

Class Activity 6G: Making up Story Problems for GCF and LCM

Class Activity 6H: Is the LCM Just the Product of the Two Numbers?

Exercises for Section 6.3 on Greatest Common Factor and Least Common Multiple

1. Calculate the least common multiple of 9 and 12 and explain why your answer is correct.

2. Calculate the greatest common factor of 36 and 63 and explain why your answer is correct.

3. Calculate the greatest common factor of 16 and 27.

4. If you have 100 pencils and 75 small notebooks, then what is the largest group of children that you can give all the pencils and all the notebooks to so that each child gets the same number of pencils and each child gets the same number of notebooks, and so that no pencils or notebooks are left over?

5. If pencils come in packages of 12 and small notebooks come in packages of 5, then how many packages of pencils and how many packages of notebooks should you buy so that you can match each pencil with a notebook?

Answers to Exercises for Section 6.3 on Greatest Common Factor and Least Common Multiple

1. The multiples of 9 are

$$9, \; 18, \; 27, \; 36, \; 45, \; 54, \; 63, \; 72, \; 81, \ldots$$

and the multiples of 12 are

$$12, \; 24, \; 36, \; 48, \; 60, \; 72, \; 84, \ldots .$$

Therefore the common multiples of 9 and 12 are

$$36, \; 72, \ldots .$$

The least of these is 36, which is therefore the least common multiple of 9 and 12.

2. The factors of 36 are

$$1, \; 2, \; 3, \; 4, \; 6, \; 9, \; 12, \; 18, \; 36$$

and the factors of 63 are

$$1, \; 3, \; 7, \; 9, \; 21, \; 63,$$

therefore the common factors of 36 and 63 are

$$1, \; 3, \; 9.$$

The greatest of these numbers is 9, therefore the greatest common factor of 36 and 63 is 9.

3. The factors of 16 are

$$1, \; 2, \; 4, \; 8, \; 16,$$

and the factors of 27 are

$$1, \; 3, \; 9, \; 27.$$

Thefore the only common factor of 16 and 27 is 1, so this must be the greatest common factor of 16 and 27.

4. If you have 100 pencils and 75 small notebooks, then what is the largest group of children that you can give all the pencils and all the notebooks to so that each child gets the same number of pencils and each child gets the same number of notebooks, and so that no pencils or notebooks are left over?

5. If pencils come in packages of 12 and small notebooks come in packages of 5, then how many packages of pencils and how many packages of notebooks should you buy so that you can match each pencil with a notebook?

Problems for Section 6.3 on Greatest Common Factor and Least Common Multiple

1. Why do we not talk about a *greatest* common multiple and a *least* common factor?

2. Make up a story problem involving concrete objects that can be solved by calculating the least common multiple of 45 and 40.

3. Make up a story problem involving concrete objects that can be solved by calculating the greatest common factor of 45 and 40.

4. In a sewing factory, seam A is sewn at a rate of 18 seams per minute, while seam B is sewn at a rate of 30 seams per minute. The factory wants to sew an equal number of seams A and seams B every minute. What are the options for how many workers should sew seam A and how many workers should sew seam B? Explain why.

6.4   Common Factors, Common Multiples, Common Denominators, and Commensurability

As we saw in Section 6.3, a common factor of two counting numbers provides a way to break the two numbers into *like parts*. We have seen this idea of breaking numbers into like parts before, namely with fractions: we find common denominators for fractions in order to work with *like parts*. To find a common denominator for a pair of fractions we simply find a common multiple of their denominators. As we'll see, like parts are related to common multiples in general. Surprisingly, even though fractions of whole numbers can always be broken into like parts, it is possible to find pairs of numbers that *can't* be broken into like parts.

Given two counting numbers, a common factor provides a way to break the two counting numbers into *like parts*. For example, Figure 6.5 shows scolls dividing a 12 cm length and a 9 cm length divided into like parts of 3 cm each.

12 cm

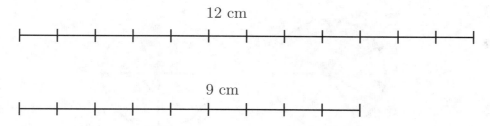

9 cm

the 12 cm and 9 cm lengths can be divided into like parts that are 3 cm long

Figure 6.5: Dividing 9 and 12 Into Like Parts

Similarly, as we saw in Sections 2.9 and 3.3, when we want to compare or add fractions, we find common denominators for the fractions so that they will both be expressed in terms of *like parts*. We can find common denominators simply by finding common multiples of the denominators. The least common multiple of the denominators will be the least common denominator of the fractions. If we think of fractions as parts of pies, as in Figure 6.6, then, when we give two fractions a common denominator by finding a common multiple of both denominators, we are really just breaking the pieces of pies into smaller, *like* pieces.

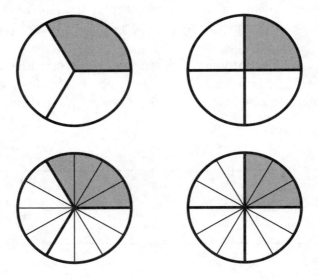

Figure 6.6: Dividing Thirds and Fourths Into Like Pieces

Notice that the way that different pieces are related dictates how they can be subdivided to make like pieces. For example, consider thirds and fourths. Let A represent a third and let B represent a fourth. Then

$$3 \times A = 4 \times B$$

because both $3 \times A$ and $4 \times B$ are equal to 1. Therefore

$$A \div 4 = B \div 3.$$

This tells us that a third should be divided into 4 equal pieces and a fourth should be divided into 3 equal pieces in order to be able to express both thirds and fourths in terms of like parts. Now let A represent 12 and let B represent 9. Once again,

$$3 \times A = 4 \times B$$

because both $3 \times A$ and $4 \times B$ are equal to 36 in this case. Therefore, once again,

$$A \div 4 = B \div 3,$$

and this common number, which is 3, is a common factor of A and B. We can think of this common factor as dividing A and B into *like parts*. In this way, common multiples are related to common factors, which yield like parts.

Class Activity 6I: Relating Common Multiples and Common Factors

Class Activity 6J: Covering Equal Areas and Dividing into Like Parts

In general, if A and B are any positive numbers and if there are *counting numbers* M and N such that

$$M \times A = N \times B,$$

then

$$A \div N = B \div M,$$

and therefore we can subdivde A and B into like parts by dividing A into N equal parts and dividing B into M equal parts. In this case, when two numbers A and B can be divided into like parts in this way, or have common multiples in this way, we say that A and B are **commensurable**.

Is every pair of positive numbers commensurable? Surprisingly, the answer is no. There are pairs of numbers that *can't* be divided into like parts. If A and B are not commensurable then we say they are **incommensurable**.

In Class Activity 6K you will see a simple way that incommensurable numbers can arise, and how the incommensurability affects pattern making.

Class Activity 6K: Pattern Tiles and Incommensurability

If you did Class Activity 6K and made patterns with pattern tiles like the ones pictured in Figure 6.7, then you probably noticed that there are often many different ways to make the same overall shape out of pattern tiles. But you may also have noticed that if a shape is made only with green triangles, thick blue rhombuses, and yellow hexagons, then, try as you might, you can't also make that same shape using orange squares or thin white rhombuses (in addition to the other shapes). This is not due to lack of cleverness on your

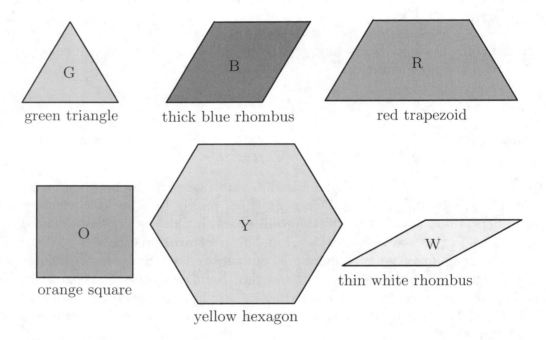

Figure 6.7: Pattern Tiles

part: it is simply impossible. Why? It is because of the kinds of numbers involved.

When we express the areas in terms of square inches, the areas of the green triangles, thick blue rhombuses, and yellow hexagons all involve the number

$$\sqrt{3},$$

whereas the areas of the orange square and the thin white rhombus are 1 and $\frac{1}{2}$. It turns out that if it were possible to make the same shape out of green triangles, thick blue rhombuses, and yellow hexagons as well as out of orange squares or thin white rhombuses in addition to the other shapes, then there would be counting numbers M and N such that

$$M\sqrt{3} = N.$$

This would mean that it would be possible to express $\sqrt{3}$ as a fraction of counting numbers, namely

$$\sqrt{3} = \frac{N}{M}.$$

However, it is possible to prove that there can be *no such counting numbers* M and N.

Exercises for Section 6.4 on Common Factors and Commensurability

1. Use the fact that
$$13167 \times 11 = 20691 \times 7$$
   to find a common factor for 13167 and 20691 (other than 1).

Answers to Exercises for Section 6.4 on Common Factors and Commensurability

1. Because
$$13167 \times 11 = 20691 \times 7,$$

   therefore
$$13167 \div 7 = 20691 \div 11.$$

   This common value of $13167 \div 7$ and $20691 \div 11$, which is 1881, is therefore a common factor of 13167 and 20691.

Problems for Section 6.4 on Common Factors and Commensurability

1. Suppose you have two types of rods, type A and type B. All type A rods are equally long and all type B rods are equally long. You find that 5 type A rods placed end to end are as long as 6 type B rods placed end to end. Explain how you can divide the type A rods and type B rods into parts so that *all* parts (from both the type A rods and the type B rods) are of equal length.

2. Suppose that you have two types of rods, type C and type D. All type C rods are equally long and all type B rods are equally long. You find that when you divide the type C rods into 6 equal parts and the type D rods into 5 equal parts, then all these parts have the same length. Describe how to place a number (a counting number) of type C rods end to end so as to be as long as a number (a counting number) of type D rods placed end to end. Explain!

6.5   Even and Odd

The study of even and odd numbers is fertile ground for investigating and exploring math. This is an area of math where it's not too hard to find interesting questions to ask about what is true, to investigate these questions, and to explain why the answers are right. Even young children can do this! In this section, in addition to asking and investigating questions about even and odd numbers, we will examine the familiar method for determining if a number is even or odd.

How would you answer the following two questions:

- What does it *mean* for a whole number to be even?

- How can you *tell* if a whole number is even?

Did you give different answers? Then why do these two different characterizations describe *the same numbers*?

The *meaning* of even is this: A whole number is **even** if it is evenly divisible by 2, in other words, if there is no remainder when you divide the number by 2. There are various other equivalent ways to say that a whole number is even:

- A whole number is even if it is evenly divisible by 2, in other words, if there is no remainder when you divide the number by 2.

- A whole number is even if it is 2 times another whole number (or another whole number times 2).

- A whole number is even if you can divide that number of things into two equal groups.

- A whole number is even if you can divide that number of things into groups of 2 with none left over (see Figure 6.8).

Figure 6.8: Even and Odd

If a whole number is not even, then we call it **odd**. Each of the ways of **odd** saying what it means for a counting number to be even described above can be modified to say what it means for a counting number to be odd:

- A whole number is odd if it is not evenly divisible by 2, in other words, if there is a remainder when you divide the number by 2.

- A whole number is odd if it is not 2 times another whole number (or another whole number times 2).

- A whole number is odd if there is one thing left over when you divide that number of things into two equal groups.

- A whole number is odd if there is one thing left over when you divide that number of things into groups of 2 with none left over (see Figure 6.8).

If you have a particular number in mind, such as 237, 921, how do you *tell* if the number is even or odd? Chances are, you don't actually divide

237,921 by 2 to see if there is a remainder or not. Instead, you look at the ones digit. In general, if a counting number has a 0, 2, 4, 6, or 8 in the ones place, then it is even; if it has a 1, 3, 5, 7, or 9 in the ones place, then it is odd. So 237,921 is odd because there is a 1 in the ones place. *But why is this a valid way to determine that there will be a remainder when you divide 237,921 by 2?* The next class activity will help you explain this.

Class Activity 6L: Why Can We Check the Ones Digit to Determine if a Number is Even or Odd?

Class Activity 6M: Problems About Even and Odd Numbers

Class Activity 6N: Questions About Even and Odd Numbers

Class Activity 6O: Extending the Definitions of Even and Odd

Exercises for Section 6.5 on Even and Odd

1. Explain why it is valid to determine if a counting number is evenly divisible by 2 by checking whether its ones digit is 0, 2, 4, 6, or 8.

2. If you add an odd number and an even number, what kind of number do you get? Explain why your answer is always correct.

3. If you multiply an even number and an even number, what kind of number do you get? Explain why your answer is always correct.

Answers to Exercises for Section 6.5 on Even and Odd

1. A counting number is evenly divisible by 2 exactly when that number of toothpicks can be divided into groups of 2 with none left over. Consider representing a counting number physically with bundled toothpicks and consider dividing the bundles of toothpicks, from the tens place on up, into groups of 2. Each bundle of 10 toothpicks can be broken into 5 groups of 2. Each bundle of 100 toothpicks can be broken into 50

groups of 2. Each bundle of 1000 toothpicks can be broken into 500 groups of 2. And so on for all the higher places. For each place from the 10s place on up, each bundle of toothpicks can always be broken evenly into groups of 2 with none left over. Therefore the full number of toothpicks can be broken into groups of 2 with none left over exactly when the number of toothpicks that are in the ones place can be broken into groups of 2 with none left over.

2. If you add an odd number to an even number, the result is always an odd number. One way to explain why is to think about putting together an odd number of blocks and an even number of blocks, as in Figure 6.9. The odd number of blocks can be divided into groups of 2 with 1 block left over. The even number of blocks can be divided into groups of 2 with none left over. When the two block collections are joined together, there will be a bunch of groups of 2, and the 1 block that was left over from the odd number of blocks will still be left over. Therefore the sum of an odd number and an even number is odd.

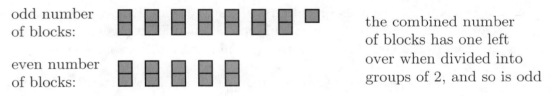

odd number of blocks:

even number of blocks:

the combined number of blocks has one left over when divided into groups of 2, and so is odd

Figure 6.9: Adding an Odd Number to an Even Number

3. If you multiply and even number and an even number, the result is always an even number. One way to explain why is to think about creating an even number of groups of blocks, where each group of blocks contains the same even number of blocks, as in Figure 6.10. According to the meaning of multiplication, the total number of blocks is the product of the number of groups with the number of blocks in each group. Because each group of blocks can be divided into groups of 2 with none left over, therefore the whole collection of blocks can be divided into groups of 2 with none left over. Therefore an even number times an even number is always an even number. (Notice that this argument actually shows that any counting number times an even number is even.)

an even number
of blocks: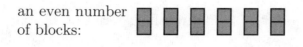

an even number of groups of an even number of blocks:

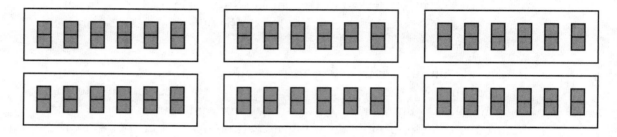

Figure 6.10: Multiplying an Even Number With an Even Number

Problems for Section 6.5 on Even and Odd

1. What is a way that the children in Mrs. Verner's Kindergarten class can tell if there are an even number or an odd number of children present in the class? Your method should not involve counting.

2. For each of the two designs in Figure 6.11, explain how you can tell whether the number of dots in the design is even or odd without determining the number of dots in the design.

3. If you add an even number and an even number, what kind of number do you get? Explain why your answer is always correct.

4. If you multiply an odd number and an odd number, what kind of number do you get? Explain why your answer is always correct.

5. If you add a number that has a remainder of 1 when it is divided by 3 to a number that has a remainder of 2 when it is divided by 3, then what is the remainder of the sum when you divide it by 3? Investigate this question by working examples. Then explain why your answer is always correct.

6. If you multiply a number that has a remainder of 1 when it is divided by 3 with a number that has a remainder of 2 when it is divided by

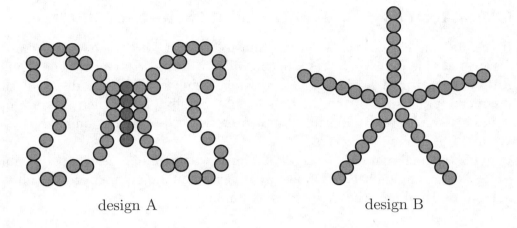

design A   design B

Figure 6.11: Two Dot Designs

3, then what is the remainder of the product when you divide it by 3? Investigate this question by working examples. Then explain why your answer is always correct.

6.6   Divisibility Tests

How can you tell if one counting number is evenly divisible by another counting number? In the previous section, we saw a way to explain why we can tell whether a counting number is even or odd by looking at its ones digit. Are there easy ways to tell if a counting number is evenly divisible by other numbers, such as 3, 4, or 5? Yes, there are, and we will study some of them in this section.

A method for *checking* or *testing* to see if a whole number is evenly divisible by another whole number is called a **divisibility test**. The **divisibility test for 2** is just the familiar test to see if a number is even or odd: check the ones digit of the number. If the ones digit is 0, 2, 4, 6, or 8, then the number is evenly divisible by 2 (it is even), otherwise the number is not evenly divisible by 2 (it is odd). This divisibility test was discussed in the previous section.

The following class activities will help you to discover the divisibility test for 3 and to explain why it is valid.

Class Activity 6P: The Divisibility Test for 3

If you did the class activity on the divisibility test for 3 then you probably discovered this divisibility test, if you did not know it already. The **divisibility test for 3** is this: given a whole number, add the digits of the number. If the sum of the digits is evenly divisible by 3, then the original number is also evenly divisible by 3, otherwise the original number is not evenly divisible by 3.

This test makes it easy to check if a large whole number is evenly divisible by 3. For example, is $127,358$ evenly divisible by 3? To answer this, all we have to do is add all the digits of this number:

$$1 + 2 + 7 + 3 + 5 + 8 = 26.$$

Because 26 is not evenly divisible by 3, therefore the original number $127,358$ is also not evenly divisible by 3. Similarly, is $111,111$ evenly divisible by 3? Because the sum of the digits is 6,

$$1 + 1 + 1 + 1 + 1 + 1 = 6,$$

and because 6 is evenly divisible by 3, therefore the original number $111,111$ must also be evenly divisible by 3.

A familiar divisibility test is the **divisibility test for 5**. It is this: given a whole number, check the ones digit of the number. If the ones digit is 0 or 5, then the original number is evenly divisible by 5, otherwise it is not evenly divisible by 5. Problem 1 will help you explain why this divisibility test is valid.

Exercises for Section 6.6 on Divisibility Tests

1. Use the divisibility test for 3 to determine which of the following numbers are evenly divisible by 3.

   (a) $125, 389, 211, 464$

   (b) $111, 111, 111, 111, 111$

   (c) $123, 123, 123, 123, 123$

   (d) $101, 101, 101, 101, 101$

2. The **divisibility test for 4** is this: given a whole number, check the number formed by its last two digits. For example, given $123,456,789$, check the number 89. If the number formed by the last two digits is evenly divisible by 4, then the original number is evenly divisible by 4, otherwise it is not. So, because 89 is not evenly divisible by 4, therefore, according to the divisibility test for 4, the number $123,456,789$ is also not evenly divisible by 4.

Explain why the divisibility test for 4 is a valid way to determine if a whole number is evenly divisible by 4.

Answers to Exercises for Section 6.6 on Divisibility Tests

1. (a) The sum of the digits of $125,389,211,464$ is $1 + 2 + 5 + 3 + 8 + 9 + 2 + 1 + 1 + 4 + 6 + 4 = 46$, which is not evenly divisible by 3, therefore $125,389,211,464$ is not evenly divisible by 3.

   (b) The sum of the digits of $111,111,111,111,111$ is 5×3, which is evenly divisible by 3, therefore $111,111,111,111,111$ is evenly divisible by 3.

   (c) The sum of the digits of $123,123,123,123,123$ is 5×6, which is evenly divisible by 3, therefore $123,123,123,123,123$ is evenly divisible by 3.

   (d) The sum of the digits of $101,101,101,101,101$ is 5×2, which is not evenly divisible by 3, therefore $101,101,101,101,101$ is not evenly divisible by 3.

2. A counting number is evenly divisible by 4 exactly when that number of toothpicks can be divided into groups of 4 with none left over. Consider representing a counting number physically with bundled toothpicks and consider dividing the bundles of toothpicks, from the hundreds place on up, into groups of 4. Each bundle of 100 toothpicks can be broken into 25 groups of 4. Each bundle of 1000 toothpicks can be broken into 250 groups of 4. Each bundle of 10000 toothpicks can be broken into 2500 groups of 4. And so on for all the higher places. For each place from the 100s place on up, each bundle of toothpicks can always be broken evenly into groups of 4 with none left over. Therefore the full number of toothpicks can be broken into groups of 4 with none left over exactly

when the number of toothpicks that are together in the tens and ones places can be broken into groups of 4 with none left over.

So, for example, if the last two digits of the original number are 25, then there will be 1 toothpick left over when the full number of toothpicks are divided into groups of 4. This is because no matter what numbers of toothpicks are in the 100s and 1000s places and so on, those can already be broken into groups of 4 with none left over. But if the last two digits of the original number are 24, then the full collection of toothpicks can be broken into groups of 4 with none left over. This is because 24 toothpicks can be divided into 6 groups of 4 with none left over, and the toothpicks in the 100s and 1000s places and so on can also be broken into groups of 4 with none left over.

Problems for Section 6.6 on Divisibility Tests

1. According to the divisibility test for 5, to check if a whole number is evenly divisible by 5 you only have to check its ones digit. If the ones digit is 0 or 5, then the number is evenly divisible by 5, otherwise it is not. Use the following ideas to help you explain why this divisibility test is a valid way to determine if a number is evenly divisible by 5.

   A whole number is evenly divisible by 5 exactly when that many tooth-picks can be divided into groups of 5 with none left over. Suppose you represent the expanded form of a whole number with bundled tooth-picks. When you divide the toothpicks into groups of 5, what will happen to each bundle of 10, each bundle of 100, each bundle of 1000, etc.? Therefore why do we only have to check the ones digit of a number to determine if it is evenly divisible by 5?

2. Beth knows the divisibility test for 3. Beth says that she can tell just by looking, and without doing any calculations at all, that the number

$$999, 888, 777, 666, 555, 444, 333, 222, 111$$

   is evenly divisible by 3. How can Beth do that? Explain why it's not just a lucky guess!

3. Sam used his calculator to calculate

$$123, 123, 123, 123, 123 \div 3.$$

Sam's calculator displayed the answer as

4.1041041041E13

Sam says that because the calculator's answer is not a whole number, therefore the number $123, 123, 123, 123$ is not evenly divisible by 3. Is Sam right? Why or why not? How do you reconcile this with Sam's calculator's display?

4. This problem will help you discover and explain the **divisibility test for 9**.

(a) Do the following at least 6 times: pick a whole number that has at least two digits, preferably more, and determine whether or not your number is evenly divisible by 9. Then add up the digits of your original whole number (before you divided by 9) and determine if this sum is divisible by 9. Compare the two results.
Example: 52371 is evenly divisible by 9 because $52371 \div 9$ is the whole number 5819, with no remainder. The sum of the digits of 52371, namely $5 + 2 + 3 + 7 + 1 = 18$, is also divisible by 9.

(b) Based on your results in part (a), formulate a *divisibility test for 9*.

(c) Use the divisibility test for 9 to determine whether or not

$139, 458, 009, 234$

is evenly divisible by 9.

(d) Now explain why the divisibility test for 9 that you formulated in part (b) is a valid way to check if a whole number is evenly divisible by 9. Use the following ideas to help you.
A whole number is evenly divisible by 9 exactly when that many toothpicks can be divided into groups of 9 with none left over. Suppose you represent the expanded form of a whole number with bundled toothpicks. Now consider dividing *each* bundle of 10, *each* bundle of 100, *each* bundle of 1000, etc. into groups of 9. Explain why the remainders give you the sum of the digits of the number.

5. (a) Is it true that a whole number is evenly divisible by 6 exactly when the sum of its digits is divisible by 6? Investigate this by considering a number of examples. State your conclusion.

(b) How could you check if the number

$$111, 222, 333, 444, 555, 666, 777, 888, 999, 000$$

is evenly divisible by 6 without using a calculator or doing long division? Explain! (*Hint:* $6 = 2 \times 3$.)

6. Investigate the questions in parts (a) and (b) by considering a number of examples.

(a) If a whole number is evenly divisible both by 6 and by 2, is it necessarily evenly divisible by 12?

(b) If a whole number is evenly divisible both by 3 and by 4 is it necessarily evenly divisible by 12?

(c) Based on your examples, what do you think the answers to the questions in (a) and (b) should be?

(d) Based on your answer to part (c), determine whether or not

$$3, 321, 297, 402, 348, 516$$

is evenly divisible by 12 without using a calculator or doing long division. Explain your reasoning. (Refer to exercise 2 on page 427.)

Bibliography

[1] Edwin Abbott Abbott. *Flatland*. Princeton University Press, 1991. See `http://eldred.ne.mediaone.net/eaa/FL.HTM`.

[2] W. S. Anglin. *Mathematics: A Concise History and Philosophy*. Springer-Verlag, 1994.

[3] Deborah Loewenberg Ball. Prospective elementary and secondary teachers' understanding of division. *Journal for Research in Mathematics Education*, 21(2):132–144, 1990.

[4] P. Barnes-Svarney, editor. *New York Public Library Science Desk Reference*. Stonesong Press, 1995.

[5] Tom Bassarear. *Mathematics for Elementary School Teachers*. Houghton Mifflin, 1997.

[6] Petr Beckmann. *A History of Pi*. St. Martin's Press, 1971.

[7] E. T. Bell. *The Development of Mathematics*. Dover, 1992. Originally published in 1945.

[8] George W. Bright. Helping elementary- and middle-grades preservice teachers understand and develop mathematical reasoning. In *Developing Mathematical Reasoning in Grades K-12*, pages 256–269. National Council of Teachers of Mathematics, 1999.

[9] California State Board of Education. *Mathematics Framework for California Public Schools*, 1999.

[10] The Carnegie Library of Pittsburgh Science and Technology Department. *Science and Technology Desk Reference*, 1993.

[11] National Research Council. *How People Learn*. National Academy Press, 1999.

[12] Frances Curcio. *Developing Data-Graph Comprehension in Grades K-8*. National Council of Teachers of Mathematics, second edition, 2001.

[13] Stanislas Dehaene. *The Number Sense*. Oxford University Press, 1997.

[14] Demi. *One Grain of Rice*. Scholastic, 1997.

[15] Tatiana Ehrenfest-Afanassjewa. *Uebungensammlung zu einer Geometrischen Propaedeuse*. Martinus Nijhoff, 1931.

[16] Euclid. *The Thirteen Books of the Elements, Translated with Commentary by Sir Thomas Heath*. Dover, 1956.

[17] National Center for Education Statistics. Findings from education and the economy: An indicators report. Technical Report NCES 97-939, National Center for Education Statistics, 1997. Available online, see http://nces.ed.gov/pubs97/97939.html.

[18] David Freedman, Robert Pisani, and Roger Purves. *Statistics*. W. W. Norton and Company, 1978.

[19] Georgia Department of Education. *Georgia Quality Core Curriculum*. See the website http://admin.doe.k12.ga.us/gadoe/sla/qcccopy.nsf.

[20] Sir Thomas Heath. *Aristarchus of Samos*. Oxford University Press, 1959.

[21] N. Herscovics and L. Linchevski. A cognitive gap between arithmetic and algebra. *Educational Studies in Mathematics*, 27:59–78, 1994.

[22] Encyclopaedia Britannica Inc. *Encyclopaedia Britannica*. Encyclopaedia Britannica, Inc., 1994. Available at www.britannica.com.

[23] H. G. Jerrard and D. B. McNeill. *Dictionary of Scientific Units*. Chapman and Hall, 1963.

[24] Graham Jones and Carol Thornton. *Data, Chance, and Probability, Grades 1 – 3 Activity Book*. Learning Resources, 1992.

[25] Constance Kamii, Barbara A. Lewis, and Sally Jones Livingston. Primary arithmetic: Children inventing their own procedures. *Arithmetic Teacher*, 41:200–203, 1993.

[26] Felix Klein. *Elementary Mathematics From An Advanced Standpoint.* Dover, 1945. Originally published in 1908.

[27] Morris Kline. *Mathematics and the Physical World.* Dover, 1981.

[28] Liora Linchevski and Drora Livneh. Structure sense: The relationship between algebraic and numerical contexts. *Educational Studies in Mathematics*, 40(2):173–196, 1999.

[29] Liping Ma. *Knowing and Teaching Elementary Mathematics.* Lawrence Erlbaum Associates, 1999.

[30] Edward Manfre, James Moser, Joanne Lobato, and Lorna Morrow. *Heath Mathematics Connections.* D. C. Heath and Company, 1994.

[31] Francis H. Moffitt and John D. Bossler. *Surveying.* Addison-Wesley, 1998.

[32] Joan Moss and Robbie Case. Developing children's understanding of the rational numbers: A new model and an experimental curriculum. *Journal for Research in Mathematics Education*, 30(2):122–147, 1999.

[33] Gary Musser and William F. Burger. *Mathematics for Elementary Teachers.* Prentice Hall, fourth edition, 1997.

[34] Phares G. O'Daffer, Randall Charles, Thomas Cooney, John Dossey, and Jane Schielack. *Mathematics for Elementary School Teachers.* Addison-Wesley, 1998.

[35] Mathematical Association of America (MAA) Committee on the Mathematical Education of Teachers. *A Call For Change: Recommendations For The Mathematical Preparation Of Teachers Of Mathematics.* The Mathematical Association Of America, 1991.

[36] U.S. Department of Education. Mathematics equals opportunity. Technical report, U.S. Department of Education, 1997. Available online, see http://www.ed.gov/pubs/math/index.html.

[37] Bureau of Labor Statistics U. S. Department of Labor. More education: Higher earnings, lower unemployment. *Occupational Outlook Quarterly*, page 40, Fall 1999. Available online, see http://stats.bls.gov/opub/ooq/ooqhome.htm.

[38] Bureau of Labor Statistics U. S. Department of Labor. Core subjects and your career. *Occupational Outlook Quarterly*, pages 26–40, Summer 1999. Available online, see http://stats.bls.gov/opub/ooq/ooqhome.htm.

[39] National Council of Teachers of Mathematics. *Curriculum and Evaluation Standards for School Mathematics*. Author, 1989.

[40] National Council of Teachers of Mathematics. *Principles and Standards for School Mathematics*. Author, 2000. See the website www.nctm.org.

[41] National Council of Teachers of Mathematics. *Navigating through Algebra in Grades 3 – 5*. National Council of Teachers of Mathematics, 2001.

[42] National Council of Teachers of Mathematics. *Navigating through Algebra in Prekindergarten – Grade 2*. National Council of Teachers of Mathematics, 2001.

[43] National Council of Teachers of Mathematics. *Navigating through Geometry in Grades 3 – 5*. National Council of Teachers of Mathematics, 2001.

[44] National Council of Teachers of Mathematics. *Navigating through Geometry in Prekindergarten – Grade 2*. National Council of Teachers of Mathematics, 2001.

[45] Conference Board of the Mathematical Sciences (CBMS). *The Mathematical Education of Teachers*, volume 11 of *CBMS Issues in Mathematics Education*. The American Mathematical Society and the Mathematical Association of America, 2001. See also http://www.cbmsweb.org/MET_Document/index.htm.

[46] The Secretary's Commission on Achieving Necessary Skills. *What Work Requires of Schools, A SCANS Report for America 2000*. U. S. Department of Labor, 1991.

[47] A. M. O'Reilley. Understanding teaching/ teaching for understanding. In Deborah Schifter, editor, *What's Happening in Math Class?*, volume 2: Reconstructing Professional Identities, pages 65–73. Teachers College Press, 1996.

[48] Dav Pilkey. *Captain Underpants and the Attack of the Talking Toilets.* Scholastic, 1999.

[49] George Polya. *How To Solve It; A New Aspect of Mathematical Method.* Princeton University Press, 1988. Reissue.

[50] Deborah Schifter. Reasoning about operations, early algebraic thinking in grades k-6. In *Developing Mathematical Reasoning in Grades K-12,* 1999 Yearbook. National Council of Teachers of Mathematics, 1999.

[51] Steven Schwartzman. *The Words of Mathematics.* The Mathematical Association of America, 1994.

[52] John Stillwell. *Mathematics and its History.* Springer-Verlag, 1989.

[53] Dina Tirosh. Enhancing prospective teachers' knowledge of children's conceptions: The case of division of fractions. *Journal for Research in Mathematics Education*, 31(1):5–25, 2000.

[54] Ron Tzur. An integrated study of children's construction of improper fractions and the teacher's role in promoting learning. *Journal for Research in Mathematics Education*, 30(4):390–416, 1999.

Index